Borel Liftings of Borel Sets: Some Decidable and Undecidable Statements

MEMOIRS
of the
American Mathematical Society

Number 876

Borel Liftings of Borel Sets: Some Decidable and Undecidable Statements

Gabriel Debs
Jean Saint Raymond

May 2007 • Volume 187 • Number 876 (first of four numbers) • ISSN 0065-9266

American Mathematical Society
Providence, Rhode Island

2000 *Mathematics Subject Classification.*
Primary 03E15; Secondary 03E45, 54H05.

Library of Congress Cataloging-in-Publication Data

Debs, Gabriel, 1952–
Borel liftings of Borel sets : some decidable and undecidable statements / Gabriel Debs, Jean Saint Raymond.
p. cm. — (Memoirs of the American Mathematical Society, ISSN 0065-9266 ; no. 876)
"May 2007, volume 187, number 876 (first of 4 numbers)."
Includes bibliographical references and index.
ISBN 978-0-8218-3971-3 (alk. paper)
1. Borel sets. 2. Constructibility (Set theory) 3. Descriptive set theory. I. Saint Raymond, Jean, 1946– II. Title.

QA248.D29 2007
511.3′22—dc22 2007060664

Memoirs of the American Mathematical Society

This journal is devoted entirely to research in pure and applied mathematics.

Subscription information. The 2007 subscription begins with volume 185 and consists of six mailings, each containing one or more numbers. Subscription prices for 2007 are US$649 list, US$519 institutional member. A late charge of 10% of the subscription price will be imposed on orders received from nonmembers after January 1 of the subscription year. Subscribers outside the United States and India must pay a postage surcharge of US$38; subscribers in India must pay a postage surcharge of US$43. Expedited delivery to destinations in North America US$53; elsewhere US$130. Each number may be ordered separately; *please specify number* when ordering an individual number. For prices and titles of recently released numbers, see the New Publications sections of the *Notices of the American Mathematical Society.*

Back number information. For back issues see the *AMS Catalog of Publications.*

Subscriptions and orders should be addressed to the American Mathematical Society, P. O. Box 845904, Boston, MA 02284-5904, USA. *All orders must be accompanied by payment.* Other correspondence should be addressed to 201 Charles Street, Providence, RI 02904-2294, USA.

Memoirs of the American Mathematical Society is published bimonthly (each volume consisting usually of more than one number) by the American Mathematical Society at 201 Charles Street, Providence, RI 02904-2294, USA. Periodicals postage paid at Providence, RI. Postmaster: Send address changes to Memoirs, American Mathematical Society, 201 Charles Street, Providence, RI 02904-2294, USA.

This publication is indexed in *Science Citation Index®*, *SciSearch®*, *Research Alert®*, *CompuMath Citation Index®*, *Current Contents®/Physical, Chemical & Earth Sciences.*
Printed in the United States of America.

∞ The paper used in this book is acid-free and falls within the guidelines established to ensure permanence and durability.
Visit the AMS home page at `http://www.ams.org/`

10 9 8 7 6 5 4 3 2 1 12 11 10 09 08 07

Contents

Abstract

One of the aims of this work is to investigate some natural properties of Borel sets which are undecidable in ZFC. Our starting point is the following elementary, though non-trivial result: Consider $X \subset 2^\omega \times 2^\omega$, set $Y = \pi(X)$, where π denotes the canonical projection of $2^\omega \times 2^\omega$ onto the first factor, and suppose that

$(\star)$: *"Any compact subset of Y is the projection of some compact subset of X".*

If moreover X is $\mathbf{\Pi}^0_2$ then (see [**1**] or [**21**])

$(\star\star)$: *"The restriction of π to some relatively closed subset of X is perfect onto Y"*

it follows (see Section 0.2) that in the present case Y is also $\mathbf{\Pi}^0_2$. Notice that the reverse implication $(\star\star) \Rightarrow (\star)$ holds trivially for any X and Y.

But as we shall see the implication $(\star) \Rightarrow (\star\star)$ for an arbitrary Borel set $X \subset 2^\omega \times 2^\omega$ is equivalent to the statement "$\forall \alpha \in \omega^\omega$, $\aleph_1$ *is inaccessible in* $L(\alpha)$". More precisely we shall prove that the validity of $(\star) \Rightarrow (\star\star)$ for all $X \in \Sigma^0_{1+\xi+1}$, is equivalent to "$\aleph_\xi^L < \aleph_1$". However we shall show independently, that when X is Borel one can, in ZFC, derive from $(\star)$ the weaker conclusion that Y is also Borel and of the same Baire class as X. This last result solves an old problem about compact covering mappings (see Section 0.2).

In fact these results are closely related to the following general boundedness principle *Lift* (X, Y): *"If any compact subset of Y admits a continuous lifting in X, then Y admits a continuous lifting in X"*, where by a lifting of $Z \subset \pi(X)$ in X we mean a mapping on Z whose graph is contained in X. The main result of this work will g ive the exact set theoretical strength of this principle depending on the descriptive complexity of X and Y. We shall also prove a similar result for a variation of *Lift* (X, Y) in which "continuous liftings" are replaced by "Borel liftings", and which answers a question of H. Friedman.

Among other applications we obtain a complete solution to a problem which goes back to Lusin concerning the existence of $\mathbf{\Pi}^1_1$ sets with all constituents in some given class $\mathbf{\Gamma}$ of Borel sets, improving earlier results by J. Stern and R. Sami.

The proof of the main result will rely on a nontrivial representation of Borel sets (in ZFC) of a new type, involving a large amount of "abstract algebra". This representation was initially developed for the purposes of this proof, but has several other applications.

Received by the editor January 30, 2004, and in revised form October 8, 2004.

2000 *Mathematics Subject Classification.* Primary 03E15; Secondary 03E45, 54H05.

Key words and phrases. Borel liftings, Luzin's Problem, Constructible universe, distinguished tree relations, compact covering .

Introduction

The interaction between regularity properties of definable sets of reals, and set-theoretical assumptions, is now a well established and familiar fact. This correspondence operates more specifically between projective classes on the one hand, and large cardinal assumptions on the other hand. Historically, the first such result is due to Solovay who proved(**[24]**) that the Perfect Set Theorem holds for all $\boldsymbol{\Pi}^1_1$ sets, if and only if for all $\alpha \in \omega^\omega$, $\aleph_1^{L(\alpha)}$ is countable (in the universe). Since then, many other similar results were discovered in connection with various natural regularity properties. Staying inside projective classes of first level, which will be the setting of this work, let us mention the equivalence, due to Martin (**[17]**) and Harrington (**[9]**), between $Det(\boldsymbol{\Sigma}^1_1)$ (determinacy of all $\boldsymbol{\Sigma}^1_1$ games) and the existence of sharps for all reals.

However, that similar phenomena might appear for classes of Borel sets is more surprising and was the starting point of this work. More precisely one can derive from the results of [**2**], [**3**],[**4**] that there exists a regularity property $\mathscr{A}$ of *pairs* of sets of reals such that $\mathscr{A}(X,Y)$ holds in ZFC whenever X is $\boldsymbol{\Pi}^0_2$ *or* Y is $\boldsymbol{\Sigma}^0_2$ and:

$$\forall X, Y \in \boldsymbol{\Pi}^1_1,\ \mathscr{A}(X,Y) \iff Det(\boldsymbol{\Sigma}^1_1)$$
$$\forall X, Y \in \boldsymbol{\Delta}^1_1,\ \mathscr{A}(X,Y) \iff \forall \alpha \in \omega^\omega, \aleph_1^{L(\alpha)} < \aleph_1$$

Moreover this property $\mathscr{A}$ is "natural" in the sense that it is not manufactured to point out some pathological behavior, but arises in a natural topological problem which was considered by topologists before our study. The statement of $\mathscr{A}$ is quite elementary but refers to several topological notions, and for more clarity we shall not explicit it now and focus for the moment on more structural aspects.

We also mention that via some codings one can derive from $\mathscr{A}$ a property $\mathscr{B}$ of sets of reals satisfying:

$$\forall Y \in \boldsymbol{\Pi}^1_1,\ \mathscr{B}(Y) \iff Det(\boldsymbol{\Sigma}^1_1)$$
$$\forall Y \in \boldsymbol{\Delta}^1_1,\ \mathscr{B}(Y) \iff \forall \alpha \in \omega^\omega, \aleph_1^{L(\alpha)} < \aleph_1$$

and which is, like "measurability" or the "Perfect Set Theorem property", an "inner regularity property", in the sense that $\mathscr{B}(Y)$ is of the form:

"If all compact subsets of Y are small then Y is small"

relatively to some smallness notion on sets of reals.

Notice that since any projective set A is constructed from a Borel set B by taking projections and complements, one can present formally any property of A as a property of B (one can even assume that B is $\boldsymbol{\Pi}^0_2$). But since a compact subset of A is not necessarily the projection of a compact subset of B, such artificial manipulations do not give rise to inner regular properties.

Lightface version: It is also an empirical observation that, at least when dealing with projective classes of the first level, all equivalences for classical regularity properties of such a class $\mathbf{\Gamma}$ admit (in general by the same arguments) "parametrized effective" versions in which the "boldface" class $\mathbf{\Gamma}$ is replaced by $\Gamma(\alpha)$ the "lightface version with parameter α" (see paragraph 0.1 below). For example if we fix $\alpha \in \omega^\omega$ then the Perfect Set Theorem holds for all $\Pi^1_1(\alpha)$ sets, if and only if $\aleph_1^{L(\alpha)}$ is countable; and $Det(\Sigma^1_1(\alpha))$ is equivalent to the existence of $\alpha^\sharp$. This is partly confirmed for property $\mathscr{A}$ since by the same arguments (letting $\alpha = 0$ for simplicity) we have:

$$\forall X, Y \in \Pi^1_1,\ \mathscr{A}(X,Y) \iff Det(\Sigma^1_1)$$

But one cannot derive from the arguments of [**6**] the corresponding equivalence for the statement "$\forall X, Y \in \Delta^1_1,\ \mathscr{A}(X,Y)$"; and as a matter of fact, it was not even clear in the present case, what the right hand side of such an equivalence would be.

The aim of this work was originally to settle this problem, which revealed to be much harder than expected. We shall indeed prove that:

$$\forall X, Y \in \Delta^1_1,\ \mathscr{A}(X,Y) \iff \aleph^L_{\omega_1^{\mathrm{CK}}} < \aleph_1$$

where ω_1^{CK} denotes the first non-recursive ordinal. In fact we shall give a level by level equivalence by proving that for any recursive ordinal ξ:

$$\forall X \in \Delta^1_1, \forall Y \in \Sigma^0_{1+\xi+1},\ \mathscr{A}(X,Y) \iff \aleph^L_\xi < \aleph_1$$

moreover we shall prove that in this last equivalence one cannot replace the class $\Sigma^0_{1+\xi+1}$ by the class $\Pi^0_{1+\xi+1}$, unless ξ is a successor. For example assuming that $\aleph^L_\omega$ is countable one can prove $\mathscr{A}(X,Y)$ for all $X \in \Delta^1_1$ and all $Y \in \Sigma^0_{\omega+1}$, but for $Y \in \Pi^0_{\omega+1}$ one needs to assume that $\aleph^L_{\omega+1}$ is countable too. In fact we shall compute, for any lightface class Γ, the exact logical strength of the statement "$\forall X \in \Delta^1_1, \forall Y \in \Gamma,\ \mathscr{A}(X,Y)$" which happens to be equivalent to "$\forall X, Y \in \Gamma,\ \mathscr{A}(X,Y)$". In this analysis the descriptive classes $D(\Sigma^0_\xi)$ and $\check{D}(\Sigma^0_\xi)$ (see section 0.1 below) will play a central role.

Both implications in the equivalences above are nontrivial. However the hard part will always be in deducing the regularity property from the set-theoretical assumption. The proof of the main result is very long and will constitute a large part of this work. One of the basic ingredients will be a general representation of Borel sets that we shall develop in the first part of this work.

Before going into more details let us fix some notations.

0.1. Descriptive classes:

All descriptive complexity computations will refer implicitly to some (equivalently any) metric compactification of the space, for example 2^ω if the space is zero-dimensional.

For any descriptive class Γ we denote by $\check{\Gamma}$ its <u>dual class</u>:

$$\check{\Gamma} := \{A^c :\ A \in \Gamma\}$$

We shall mainly deal with the projective classes $\mathbf{\Sigma}^1_1$, $\mathbf{\Pi}^1_1$, though some results extend trivially to the class $\mathbf{\Sigma}^1_2$. We shall also consider various classical subclasses of the class $\mathbf{\Delta}^1_1$ of all Borel sets namely: the additive and multiplicative <u>*Baire classes*</u> $\mathbf{\Sigma}^0_\xi$ and $\mathbf{\Pi}^0_\xi$, but also the <u>*Lavrentieff differences classes*</u> $D_\eta(\mathbf{\Sigma}^0_\xi)$ as well as their dual classes $\check{D}_\eta(\mathbf{\Sigma}^0_\xi)$.

We recall that if η and ξ are countable ordinals, a set A is said to be $D_\eta(\mathbf{\Sigma}^0_\xi)$ if there exists a transfinite family $(A_\zeta)_{\zeta<\eta}$ of $\mathbf{\Sigma}^0_\xi$ sets such that:

$$x \in A \iff \text{the least } \zeta < \eta \text{ such that } x \in A_\zeta \text{ is of parity opposite to } \eta$$

and A is said to be $\check{D}_\eta(\mathbf{\Sigma}^0_\xi)$ if its complement is $D_\eta(\mathbf{\Sigma}^0_\xi)$.

In particular $D_1(\mathbf{\Sigma}^0_\xi) = \mathbf{\Sigma}^0_\xi$ and $\check{D}_1(\mathbf{\Sigma}^0_\xi) = \mathbf{\Pi}^0_\xi$, and for $\eta = 2$ we shall denote more simply by $D(\mathbf{\Sigma}^0_\xi)$ and $\check{D}(\mathbf{\Sigma}^0_\xi)$ the classes $D_2(\mathbf{\Sigma}^0_\xi)$ and $\check{D}_2(\mathbf{\Sigma}^0_\xi)$. Notice that:

$$D(\mathbf{\Sigma}^0_\xi) = \{A \cap B \;:\; A \in \mathbf{\Sigma}^0_\xi \,,\, B \in \mathbf{\Pi}^0_\xi\}$$
$$\check{D}(\mathbf{\Sigma}^0_\xi) = \{A \cup B \;:\; A \in \mathbf{\Sigma}^0_\xi \,,\, B \in \mathbf{\Pi}^0_\xi\}$$

Finally we recall that by a classical result of Hausdorff-Kuratowski([**11**]) the *ambiguous Baire classes* are described in the successor case by:

$$\mathbf{\Delta}^0_{\xi+1} = \bigcup_{\eta<\omega_1} D_\eta(\mathbf{\Sigma}^0_\xi)$$

In the context of classical spaces such as 2^ω or ω^ω we shall also consider the *"lightface"* classes:

$$\Sigma^1_1,\, \Pi^1_1,\, \Delta^1_1,\, \Sigma^0_\xi \,,\, \Pi^0_\xi \,,\, D_\eta(\Sigma^0_\xi),\, \check{D}_\eta(\Sigma^0_\xi)$$

where the ordinals η and ξ are then implicitly supposed to be recursive. We shall refer to these classes as the *"effective versions"* of the *"boldface"* classes:

$$\mathbf{\Sigma}^1_1,\, \mathbf{\Pi}^1_1,\, \mathbf{\Delta}^1_1,\, \mathbf{\Sigma}^0_\xi \,,\, \mathbf{\Pi}^0_\xi \,,\, D_\eta(\mathbf{\Sigma}^0_\xi),\, \check{D}_\eta(\mathbf{\Sigma}^0_\xi)$$

If Γ is the effective version of $\mathbf{\Gamma}$ then for any $\alpha \in \omega^\omega$ one can also consider $\Gamma(\alpha)$ the *"effective version of* $\mathbf{\Gamma}$ *with parameter* α*"*. We recall that for all the classes considered here we have the following parametrization result:

$$\mathbf{\Gamma} = \bigcup_{\alpha\in\omega^\omega} \Gamma(\alpha)$$

Throughout the paper by <u>class</u> (without any additional precision) we will mean one of the "lightface" or "boldface" classes above. In particular *a class of Borel sets will implicitly be a Lavrentieff class.*

However in a very few number of sections we shall also discuss briefly the extension of some of the results to *Wadge classes*; but this more general context is marginal in our study and will then be explicitly indicated.

For all undefined notions and other basic properties of classical descriptive classes we refer the reader to [**11**] or[**18**].

0.2. An elementary topological problem:

To present the problem which initially motivated this work, we need to introduce a number of topological notions. However we can put the reader's mind at ease immediately by saying that none of these notions will really be needed for the sequel. In fact we shall, in a preliminary step, give in a purely descriptive language a more intuitive reformulation of the original topological problem.

In all the sequel $f : X \to Y$ will denote a continuous and onto mapping between two separable and metrizable spaces.

<u>Perfect mappings</u>:

Recall that the mapping f is said to be *perfect* if *the inverse image by f of any compact subset of Y is a compact subset of X.*

Obviously any continuous onto mapping with compact domain is perfect. More generally starting from a continuous onto mapping $f : K \to L$ with compact domain, for any subset $Y \subset L$ the restriction of f to $X := f^{-1}(Y)$ is a perfect mapping from X onto Y.

Perfect mappings play an important role in various areas of Analysis. But here we shall mainly be concerned with their descriptive properties. More particularly we point the following fundamental property (see for example [**22**]):

The perfect image of a Borel space is also Borel.

By this we mean that if $f : X \to Y$ is perfect and onto and X is Borel, then Y is Borel (but of course the image by f of a Borel subset of X will not be necessarily Borel). Moreover if X is $\mathbf{\Sigma}^0_\xi$ or $\mathbf{\Pi}^0_\xi$ with $\xi \geq 2$ then Y is of the same Baire class; in particular the perfect image of a Polish space is a Polish space.

<u>Inductively perfect mappings</u>:

The mapping f is said to be *inductively perfect* if *there exists a subset X' of X such that $f(X') = Y$ and the restriction mapping $f_{|X'} : X' \to Y$ is perfect.* It is easily seen that such a subset X' is necessarily closed *relatively* to X; in particular if X is Borel, then X' is also Borel and of the same Baire class. Hence:

The inductively perfect image of a Borel space is also Borel.

<u>Compact covering mappings</u>:

The mapping f is said to be *compact covering* if *any compact subset of Y is the direct image of some compact subset of X.* Thus one evidently has:

$$\text{“}f \text{ perfect”} \implies \text{“}f \text{ inductively perfect”} \implies \text{“}f \text{ compact covering”}$$

and it is quite clear that the first implication is strict. Also, using the Axiom of Choice (more precisely the existence of Bernstein sets) it is not difficult to construct counter-examples for the converse of the second implication. However for “nice” spaces this converse does hold:

a) Any compact covering mapping with Polish domain is inductively perfect. (In particular the compact covering image of a Polish space is a Polish space)

b) Any compact covering mapping with σ-compact range is inductively perfect.

a) was proved independently by Christensen ([**1**]) and Saint Raymond ([**21**]), and *b)* by Ostrovsky ([**20**]) (also by Just and Wicke ([**10**]) for mappings with countable range). In view of these results the following question arises naturally:

<u>Problem I</u>: *Is any compact covering mapping between two Borel spaces inductively perfect?*

In fact it follows from [**2**] and [**3**] that the answer is positive if *Det*($\mathbf{\Sigma}^1_1$) holds, but negative in L even for very simple Borel (for example $\mathbf{\Delta}^0_3$) spaces.

We can now introduce the regularity property $\mathscr{A}$ considered at the beginning. For any spaces X and Y consider the statement:

$\mathscr{A}(X,Y)$: *Any compact covering mapping from X onto Y is inductively perfect.*

Also given two classes Λ and Γ we set:

$\mathscr{A}(\Lambda,\Gamma)$: *Any compact covering mapping from $X \in \Lambda$ onto $Y \in \Gamma$ is inductively perfect.*

Thus we are interested in the status of $\mathscr{A}(\Lambda,\Gamma)$ for various classes of Borel sets. However we point out that it follows from the arguments of ([**3**], Theorem 6.5) that for any class Γ such that $\check{D}(\Sigma^0_2) \subset \Gamma \subset \Delta^1_1$ we have:

$$\mathscr{A}(\Gamma,\Gamma) \iff \mathscr{A}(\Delta^1_1,\Gamma) \iff \mathscr{A}(\Pi^1_1,\Gamma)$$

and the same holds for "boldface" classes. So this regularity property of Borel sets hides, but in a natural way, a regularity property of $\mathbf{\Pi}^1_1$ sets.

To finish let us mention that the study of property $\mathscr{A}$ for arbitrary (separable metrizable) spaces can be reduced to <u>projection mappings</u> inside zero-dimensional spaces, that is mappings $\pi_{|X}$ which are the restrictions of the canonical projection $\pi : 2^\omega \times 2^\omega \to 2^\omega$ on the first factor to some set $X \subset 2^\omega \times 2^\omega$. More precisely given two classes Λ and Γ consider the statement:

$$\mathbb{A}(\Lambda,\Gamma) := \begin{cases} \text{“\textit{For any } } X \subset 2^\omega \times 2^\omega \text{ \textit{in} } \Lambda \text{ \textit{and any} } Y \subset 2^\omega \text{ \textit{in} } \Gamma, \text{ \textit{if} } \pi_{|X} \text{ \textit{is}} \\ \text{\textit{compact covering on} } Y \text{ \textit{then} } \pi_{|X} \text{ \textit{is inductively perfect on} } Y\text{”} \end{cases}$$

then by ([**2**], Theorem 3.1):

$$\mathscr{A}(\Lambda,\Gamma) \iff \mathbb{A}(\Lambda,\Gamma)$$

0.3. Continuous and Borel liftings:

We discusss now the descriptive reformulation of property $\mathscr{A}$ that we will really handle in the proofs. In fact we shall introduce a formally stronger property which, *in the presence of some appropriate set-theoretical assumptions*, will be equivalent to $\mathscr{A}$.

Recall that a <u>section</u> for a mapping $f : X \to Y$ is a mapping $\varphi : Y \to X$ such that the composition $f \circ \varphi$ is the identity on Y. More generally given any subset Y' of Y, a section of f *on* Y' is a mapping $\varphi : Y' \to X$ such that the composition $f \circ \varphi$ is the identity on Y'.

Any perfect mapping $f : X \to Y$ admits a first Baire class section (for example if $X \subset 2^\omega$ one can define $\varphi(y)$ to be the lexicographical minimum of the compact set $f^{-1}(y)$). Hence any inductively perfect mapping $f : X \to Y$ admits a first Baire class section too.

Conversely if $f : X \to Y$ admits a continuous section φ then the restriction of f to $\varphi(X)$ is clearly perfect, hence f is inductively perfect. On the other hand noting that any (continuous onto) mapping on a compact space is automatically perfect we get:

$$\begin{array}{ccc}
f \text{ admits a continuous section} & \Rightarrow & \begin{cases} f \text{ admits a continuous section} \\ \text{on any compact subset of } Y \end{cases} \\
\Downarrow & & \Downarrow \\
f \text{ is inductively perfect} & \Rightarrow & f \text{ is compact covering} \\
\Downarrow & & \Downarrow \\
f \text{ admits a first Baire class section} & \Rightarrow & \begin{cases} f \text{ admits a first Baire class section} \\ \text{on any compact subset of } Y \end{cases}
\end{array}$$

Then in connection to Problem I one can naturally consider the following inner regularity problem.

Problem II: *Let Φ be a class of functions, and $f : X \to Y$ be a continuous onto mapping. If for any compact subset L of Y the restriction of f to $f^{-1}(L)$ onto L admits a section in Φ, does f admit a section in Φ ?*

More particularly we will be interested in the case where Φ is the class of all continuous (first Baire class, Baire class ξ, Borel) mappings. Notice that a priori one cannot derive from the previous diagram any trivial connection between Problem I and Problem II. But as for Problem I one can also reduce Problem II to the particular case of projection mappings, for which we shall introduce now a more specific terminology.

Let $\pi_{|X} : X \to Y$ be a projection mapping (recall that $X \subset 2^\omega \times 2^\omega$, $Y \subset 2^\omega$, and π is the projection on the first factor). If $Y' \subset Y$ we shall say that a mapping $\varphi : Y' \to 2^\omega$ is a lifting of Y' in X if the mapping $\tilde{\varphi} : Y' \to 2^\omega \times 2^\omega$ defined by $\tilde{\varphi}(y) = (y, \varphi(y))$ is a section of $\pi_{|X}$ on Y'. Then for any classes Λ and Γ we consider the following statements:

$$\mathit{Lift}(\Lambda,\Gamma) := \begin{cases} \text{“For all } X \subset 2^\omega \times 2^\omega \text{ in } \Lambda \text{ and } Y \subset 2^\omega \text{ in } \Gamma, \text{ if any compact} \\ \text{subset of } Y \text{ has a continuous lifting in } X \text{ then } Y \text{ has a} \\ \text{continuous lifting in } X\text{”} \end{cases}$$

$$\mathit{Lift}^{(\xi)}(\Lambda,\Gamma) := \begin{cases} \text{“For all } X \subset 2^\omega \times 2^\omega \text{ in } \Lambda \text{ and } Y \subset 2^\omega \text{ in } \Gamma, \text{ if any compact} \\ \text{subset of } Y \text{ has a Borel lifting of class } \xi \text{ in } X \text{ then } Y \text{ has} \\ \text{a Borel lifting of class } \xi \text{ in } X\text{”} \end{cases}$$

$$\mathit{Lift}^{*}(\Lambda,\Gamma) := \begin{cases} \text{“For all } X \subset 2^\omega \times 2^\omega \text{ in } \Lambda \text{ and } Y \subset 2^\omega \text{ in } \Gamma, \text{ if any compact} \\ \text{subset of } Y \text{ has a Borel lifting in } X \text{ then } Y \text{ has a Borel} \\ \text{lifting in } X\text{”} \end{cases}$$

Notice that even when the classes Λ and Γ are “lightface” we do not require in the previous statements, any “lightface” property for the liftings.

Again formally speaking these statements are not a priori comparable to $\mathbb{A}(\Lambda,\Gamma)$. However for $\Gamma \subset \Lambda = \Pi^1_1$ we shall show that:

$$\begin{array}{ccccc}
\mathit{Lift}(\Pi^1_1,\Gamma) & \Longrightarrow & \mathit{Lift}^{(\xi)}(\Pi^1_1,\Gamma) & \Longrightarrow & \mathit{Lift}^{(*)}(\Pi^1_1,\Gamma) \\
\mathit{Lift}(\Pi^1_1,\Gamma) & \Longrightarrow & \mathbb{A}(\Pi^1_1,\Gamma) & \Longrightarrow & \mathit{Lift}^{(*)}(\Pi^1_1,\Gamma)
\end{array}$$

and it will follow from the main result that for $\Gamma \subset \Delta^1_1$, $\mathit{Lift}(\Pi^1_1, \Gamma)$ and $\mathbb{A}(\Pi^1_1, \Gamma)$ are actually equivalent, while as we shall see $\mathit{Lift}^{(\star)}(\Pi^1_1, \Gamma)$ is much weaker.

More precisely we shall prove that

$$\mathit{Lift}(\Pi^1_1, \Delta^1_1) \iff \aleph^L_{\omega_1^{\mathrm{CK}}} < \aleph_1$$

and if we denote by $\leq^*$ the partial order of eventual domination on ω^ω then

$$\mathit{Lift}^{(\star)}(\Pi^1_1, \Delta^1_1) \iff \omega^\omega \cap L \textit{ is } \leq^*\textit{-bounded in } \omega^\omega$$

This last equivalence answers a question of H. Friedman ([**8**]).

Finally, we point out that unlike for $\mathbb{A}$, $\mathit{Lift}(\Pi^1_1, \Gamma)$ is *not* equivalent to $\mathit{Lift}(\Delta^1_1, \Gamma)$, and as a matter of fact $\mathit{Lift}(\Delta^1_1, \Delta^1_1)$ is provable in ZFC.

0.4. Main result

A careful analysis of the proofs of Theorem 3.2 and Theorem 9.10 in [**6**], shows that for any recursive ordinal ξ, $\mathit{Lift}(\Pi^1_1, \Sigma^0_{1+\xi+1})$ can be derived from "$\aleph^L_\eta < \aleph_1$" with $\omega^\xi \leq \eta < \omega_1^{\mathrm{CK}}$, and the main result of this work will determine the largest class Γ_ξ for which $\mathit{Lift}(\Pi^1_1, \Gamma_\xi)$ can be derived from "$\aleph^L_\xi < \aleph_1$". This class happens to be neither an additive nor a multiplicative Baire class in general, and depends on whether ξ is limit or successor. More precisely for any recursive ordinal ξ if we set:

$$\Gamma_\xi = \begin{cases} \Sigma^0_{1+\xi+1} & \text{if } \xi \text{ is limit} \\ D(\Sigma^0_{1+\xi+1}) & \text{if } \xi \text{ is successor} \end{cases}$$

then

THEOREM A. *If $\aleph^L_\xi < \aleph_1$ then $\mathit{Lift}(\Pi^1_1, \Gamma_\xi)$.*

The proof of Theorem A will combine the elaborated arguments of ([**6**], Theorem 3.2) with the "double-tree representation" for Borel sets that we will develop in the first part of this work. This result is best possible, and we shall prove the following strong converse.

THEOREM B. *If $\mathit{Lift}(\Pi^1_1, \check{\Gamma}_\xi)$ then $\aleph^L_{\xi+1} < \aleph_1$.*

0.5. Application 1

In his monograph on analytic sets Lusin asks explicitly ([**16**]], p. 207) the following question:

Problem III (Lusin): *Does there exist a $\mathbf{\Pi}^1_1$ non-Borel set with all constituents of bounded Borel rank ?*

This problem is considered by Lusin as a "restricted" form of the more general problem, that he attributes to Lebesgue, of the existence of a "definable" subset of the real line of cardinality $\aleph_1$. Of course at the time Lusin formulates his question a constituent of $\mathbf{\Pi}^1_1$ set C was to be understood in the classical sense as the set of all elements of C of constant index relatively to some fixed Souslin representaion of the complement of C.

More generally given a $\mathbf{\Pi}^1_1$-norm $\varphi : C \to \omega_1$ on a $\mathbf{\Pi}^1_1$ non Borel set C one can ask whether all φ-constituents of C (i.e. the sets of the form $\{\varphi = \mu\}$ for some $\mu < \omega_1$) can be of bounded Borel rank. One can also consider the similar question for φ-initial segments of C (*i.e.* the sets of the form $\{\varphi \leq \mu\}$); but since constituents and initial segments are related by fixed Boolean operations, these two problems are obviously equivalent.

Even with this modern reformulation Lusin's problem was in fact completely settled by the works of Stern ([**25**], [**26**]) and Sami ([**23**]), from which it follows that the answer is negative if and only if "$\forall\,\alpha,\ \aleph_1^{L(\alpha)} < \aleph_1$". But if we *fix* some class $\mathbf{\Gamma} \subset \mathbf{\Delta}_1^1$ and consider the statements

$$\mathscr{N}_{\{\leq\}}(\Pi_1^1, \mathbf{\Gamma}) : \text{"Any } \Pi_1^1 \text{ set admitting a } \Pi_1^1\text{-norm with all initial segments in } \mathbf{\Gamma} \text{ is in } \mathbf{\Gamma}\text{"}$$

$$\mathscr{N}_{\{=\}}(\Pi_1^1, \mathbf{\Gamma}) : \text{"Any } \Pi_1^1 \text{ set admitting a } \Pi_1^1\text{-norm with all constituents in } \mathbf{\Gamma} \text{ is in } \mathbf{\Gamma}\text{"}$$

then firstly these two statements are neither equivalent, nor comparable anymore (unless $\mathbf{\Gamma}$ is an additive Baire class), and secondly it is not clear from Stern's and Sami's arguments, what is the precise "amount" of "$\forall\,\alpha,\ \aleph_1^{L(\alpha)} < \aleph_1$" that one really needs to ensure one or the other of these statements. Basing on Theorems A and B above we shall again compute the best possible class $\mathbf{\Gamma}$ that one can treat under the assumption "$\aleph_\xi^L < \aleph_1$". This leads us to consider new boundary classes by setting for all ξ:

$$\begin{cases} \mathbf{\Sigma}_\xi^* = \mathbf{\Sigma}_{1+\xi}^0 \text{ and } \mathbf{\Pi}_\xi^* = \mathbf{\Pi}_{1+\xi}^0 & \text{if } \xi \text{ is limit} \\ \mathbf{\Sigma}_\xi^* = \mathbf{\Sigma}_{1+\xi+1}^0 \text{ and } \mathbf{\Pi}_\xi^* = \mathbf{\Pi}_{1+\xi+1}^0 & \text{if } \xi \text{ is successor} \end{cases}$$

Notice that $\mathbf{\Sigma}_\xi^*$ is the largest additive Baire class contained in $\check{\mathbf{\Gamma}}_\xi$

THEOREM C. *Let ξ be any recursive ordinal.*

a) If $\aleph_\xi^L < \aleph_1$ then $\mathscr{N}_{\{\leq\}}(\Pi_1^1, \mathbf{\Gamma}_\xi)$ and $\mathscr{N}_{\{=\}}(\Pi_1^1, \mathbf{\Sigma}_\xi^*)$.

b) If $\mathscr{N}_{\{\leq\}}(\Pi_1^1, \check{\mathbf{\Gamma}}_\xi)$ or $\mathscr{N}_{\{\leq\}}(\Pi_1^1, \mathbf{\Pi}_\xi^*)$ then $\aleph_{\xi+1}^L < \aleph_1$.

We shall also prove some new Perfect Set Theorems for equivalence relations with Borel equivalence classes of bounded rank, similar to earlier results by Stern ([**25**], [**26**]).

0.6. Application 2

Since the image of a Borel space by an inductively perfect mapping is Borel, one can very naturally consider the following weaker form of Problem I:

Problem IV: (Ostrovsky) *Is the compact covering image of a Borel space also Borel ?*

This old problem was raised by the early results of [**1**] and [**21**] where the $\mathbf{\Pi}_2^0$ case is proved, and reformulated more recently by Ostrovsky in [**19**]. We shall give, in ZFC, a positive answer to Problem IV. In fact we shall derive the solution from the following more general result:

THEOREM D. *Let $f : X \to Y$ be Borel mapping with Borel domain, and ξ a countable ordinal. If any compact subset of Y admits a Borel section of class ξ, then Y is Borel and admits a Borel section of class ξ.*

COROLLARY E. *If $f : X \to Y$ is compact covering and X is Borel then Y is also Borel and of the same Baire class as X.*

Observe that the statement of Theorem D does not involve any set-theoretical assumption. Its proof does not rely on Theorem A but will make a crucial use, though in a completely different way, of the "double-tree representation" of Borel sets which is one of the fundamental ingredients of the proof of Theorem A. The last part of the conclusion in Corollary E is based on earlier results from [**2**].

0.7. Application 3

From some technical variations of Theorems A, B and Corollary E, one can derive the following result which provides a full solution to Problem I. For simplicity we restrict the statement to projection mappings $\pi_X : X \to Y$ (see 0.2).

THEOREM F. *For any recursive ordinal ξ the following are equivalent:*

(1) $\aleph_\xi^L < \aleph_1$
(2) *Any compact covering projection from a Π_1^1 set onto a Γ_ξ set is inductively perfect.*
(3) *Any compact covering projection from a Γ_ξ set onto a Γ_ξ set is inductively perfect.*
(4) *Any compact covering projection from a Γ_ξ set is inductively perfect.*

The paper can be split in three parts, of unequal volume: in the first part (Chapters 1 and 2) we discuss the general double-tree representation of Borel sets mentioned above. The applications we give of this representation are of two completely different types and constitute the core of the two remaining parts which are mutually independent. The first type of application is developed in Chapter 3 where we prove a preliminary version of Theorem D. In this chapter we also give a new proof of a Hurewicz type result due to Louveau and Saint Raymond ([**14**]). In Chapter 4 we give the complete solution to Problem IV, and also answer a question of H. Friedman on Borel liftings. The second type of application of the double-tree representation of Borel sets is provided by the proof of Theorem A which constitutes the central part of this work. The proof of this result is very long and will be given in Chapter 6. In Chapter 5 we discuss various consequences of Theorem A and prove the strong converse given by Theorem B; the arguments are completely independent from the other parts of this work and this chapter can be read separately.

Thus aside an obvious linear approach the paper can be read by following the paths of the diagram below:

$$\begin{array}{ccccccc} & & 3 & \longrightarrow & 4 & & \\ & \nearrow & & & & & \\ 1 & & & & & & 5 \\ & \searrow & & & & \nearrow & \\ & & 2 & \longrightarrow & 6 & & \end{array}$$

Acknowledgement: We are grateful to Ramez L. Sami for long discussions and explanations concerning [**23**] and which were valuable to elaborate the new developments given in Section 5.2.

We would also like to thank the referee for his careful reading of the paper and his numerous comments which helped us to improve the presentation of the final version.

CHAPTER 1

A Tree Representation for Borel Sets

In this chapter we shall develop a representation of Borel sets, which is quite new in its spirit. We point out that in this representation *tree relations* will play a central role by coding continuous *mappings* and not merely closed *sets* as it is commonly the case.

In fact one can very naturally represent a Borel set A by coding in one way or another the pointset operations which construct A from the basic open sets; this leads to the standard notion of "Borel code" that we shall not use here. To explain the main idea behind our representation let us recall the following classical result of Lusin: any Borel set A is the range of a one-to-one continuous mapping $\varphi : F \to A$ where F is a closed subset of ω^ω. Notice that in this case, by another result of Lusin, the inverse mapping $\varphi^{-1} : A \to F$ is automatically Borel; and by a result of Kuratowski, one can ensure moreover that the Baire class of φ^{-1} is essentially equal to the Borel rank of A. All these results are quite elementary.

Our representation is in some sense a finite version of Kuratowski's representation described above, in which the mapping φ is of a specific type and can be very naturally derived from some tree relation. More precisely a Borel subset A of ω^ω will be represented by a transfinite family $(R^{(\eta)})_{\eta\leq\xi}$ of tree relations *on* $\omega^{<\omega}$ (so $R^{(\eta)} \subset \omega^{<\omega} \times \omega^{<\omega}$) with ξ essentially equal to the Borel rank of A.

The main goal of the chapter is to introduce the basic concepts needed to state and then prove this representation Theorem. This proof is quite long and is based on a number of elaborate abstract constructions, that we will describe in the first part of the chapter. More particularly the treatment of Borel sets of limit rank λ is delicate, and one has to distinguish the case where the limit ordinal is additively decomposable (such as $\lambda = \omega + \omega$) or not (such as $\lambda = \omega^2$).

1.1. Trees

1.1.1. Sequential trees. For any set A we denote by Seq(A) the set $A^{<\omega}$ of all finite sequences in A, and by $\prec$ the standard *(strict)* extension relation on Seq(A). But we shall also need to use the *(non strict)* extension relation $\preceq$ that we shall most often denote by Ext, so:

$$u \; Ext \; v \iff (u \prec v \text{ or } u = v)$$

We shall say that a subset T of Seq(A) is a <u>sequential tree</u> on A if it is left hereditary relatively to $\prec$ (equivalently to *Ext*), *i.e.*: if $v \in T$ and $u \prec v$ then $u \in T$.

If T is a sequential tree on a set A we denote as usual by $\lceil T \rceil$ the set of all *infinite branches* of T that is the set:

$$\lceil T \rceil = \{\alpha \in A^\omega : \quad \forall n \in \omega, \; \alpha_{|n} \in T\}$$

It is a basic elementary fact that sets of this form are exactly the closed subsets of the product space A^ω (where A is equipped with the discrete topology).

We assume that the reader is familiar with classical classes of κ-Souslin sets. We shall namely use frequently the standard representation of $\mathbf{\Sigma}^1_1$, $\mathbf{\Pi}^1_1$ sets as the projections of sets of the form $\lceil T \rceil$ where T is a sequential tree on $\omega \times \omega$, $\omega \times \omega_1$ (see [**11**] or [**18**] for more details).

1.1.2. Warning. It is more common to refer to a sequential tree simply as a "tree". However we shall not follow this terminology in this work, and use always the term "sequential tree". Even more we shall use the term "tree" as a shortening for "tree relation". We shall justify later on these choices of terminology which are not arbitrary.

1.1.3. Tree relations. A *tree relation* or more simply a *tree* R on a set E is a partial order relation on E with a least element, that we shall always denote by 0, and such that the set of all predecessors of a fixed element in E is finite and linearly ordered by R.

Notice that if R and S are tree relations on E with $R \subset S$ then they necessarily have the same least element.

The fundamental example of a tree relation is provided by the relation *Ext* on some sequential space.

By *a predecessor* of an element $a \in E$ we shall always mean a "strict" predecessor that is an element $a' \neq a$ such that $a' \; R \; a$.

For any $a \in E$ the set of all predecessors of a admits a largest element to which we shall refer as *the predecessor* of a and the mapping $\varphi : E \to E$ defined by:

$$\varphi(a) = \begin{cases} 0 & \text{if } a = 0 \\ \textit{the } R\textit{-predecessor of } a & \text{if } a \neq 0 \end{cases}$$

will be called *the predecessor function* of R.

If $\sigma_R(a) = (a_0, a_1, \ldots, a_{n-1}) \in \mathrm{Seq}(E)$ is the unique increasing enumeration of the set of all predecessors of a, the integer $n = |\sigma_R(a)|$, that is the number of (strict) predecessors of a, is called the *height* of a and is denoted by $h_R(a)$; in particular $h_R(0) = 0$.

We recall some basic notions and properties of tree relations: If R is a tree relation on E, an R-*chain* is a subset of E which is totally ordered by R. An R-chain A admits a unique enumeration for which the height function h_R is monotone; we shall refer to this enumeration as *the R-enumeration* of A. We shall very often identify an R-chain $A \subset E$ with its R-enumeration $\alpha \in E^{<\omega} \cup E^\omega$. The sequence $\sigma_R(a)$ defined above will be called the *strict R-branch* or *strict R-chain* of a; we shall also consider the sequence $\bar{\sigma}_R(a) := \sigma_R(a)^\frown\langle a \rangle$ that we will call the *R-branch* or *R-chain* of a.

An R-*branch* is a maximal (relatively to the inclusion relation $\subseteq$) R-chain. Any *infinite* chain is contained in a *unique* infinite branch. If A is an *infinite* R-branch then the restriction to A of the height function h_R defines a bijection from A *onto* ω, whose inverse is precisely the R-enumeration of A.

We shall denote by $[R]$ the set of all infinite R-branches, that we again shall view (via the R-enumerations) as a subset of E^ω, so:

$$[R] = \{\, \alpha \in E^\omega \; : \quad \forall n \in \omega, \;\; \alpha(n) \; R \; \alpha(n+1) \;\; \text{and} \;\; h_R(\alpha(n)) = n \,\}$$

We also equip $[R]$ with the topology induced by the product of the discrete topology on E. A basis for the topology of $[R]$ is hence given by the basic clopen sets :

$$N_a^R = \{ \alpha \in [R] \ : \quad \alpha(n) = a \}$$

where $a \in E$ and $n = h_R(a)$. Clearly $[R]$ is a closed subset of E^ω, hence the topology of $[R]$ can be defined by a complete metric. In particular if E is countable, which will always be the case in the sequel, then $[R]$ is a Polish space, which can be identified with a closed subspace of ω^ω.

1.1.4. Tree relations versus sequential trees. It is an obvious fact that tree relations are in a canonical one-to-one correspondence with sequential trees: if T is a sequential tree on the set A then $\mathit{Ext}(T) := \mathit{Ext} \cap (T \times T)$ is a tree relation with domain T; moreover the height of any element $t \in T$ is precisely the length of the sequence t. Conversely if R is a tree relation on a set E then $T_R = \{\sigma_R(a);\ a \in E\}$ is a sequential tree on E, and σ_R defines an isomorphism between the structures (E, R) and $(T_R, \mathit{Ext}(T_R))$. However, depending on the context, one or the other notion might be more suitable.

For example, as we mentioned in the introduction of this chapter, we shall associate to any Borel set in ω^ω a transfinite family $(R^{(\eta)})_{\eta \leq \xi}$ of tree relations *on* $\mathrm{Seq}(\omega)$; and if we want to stick to sequential trees, we should identify each $R^{(\eta)}$ to a sequential tree $T^{(\eta)}$ *on* $\mathrm{Seq}(\omega)$, so $T^{(\eta)} \subset \mathrm{Seq}(\mathrm{Seq}(\omega)) = \omega^{<\,(\omega^{<\omega})}$. It is clear that such identifications would have a very negative effect on the readability of the paper.

On the other hand many of the basic concepts from Descriptive Set Theory which we will use very frequently, such as games or classical representations of $\mathbf{\Sigma}_1^1$ or $\mathbf{\Pi}_1^1$ sets, involve naturally sequential trees and not tree relations, and relating them in the language of tree relations would be not only artificial but quite misleading for the reader.

For these reasons we shall use in the sequel both notions of tree relations and sequential trees. In particular we shall very often identify a classical sequential space A^ω, such as the Baire space ω^ω or the Cantor space 2^ω, to the corresponding space $[\mathit{Ext}]$, by identifying an element $\alpha \in A^\omega$ with the infinite branch $(\alpha_{|n})_{n\in\omega}$ of $(\mathrm{Seq}(A), \mathit{Ext})$. However we shall operate these identifications only inside the proofs and state all the results in the more natural context of sequential spaces.

From now on, *E will denote a countable set, and S will be a fixed tree relation on E.* In fact for all applications we will take $(E, S) = (\mathrm{Seq}(A), \mathit{Ext})$; however in many inductive proofs one needs to consider arbitrary tree relations.

1.2. Distinguished subrelation

DEFINITION 1.2.1. Let S be a tree relation. A subrelation $R \subset S$ is said to be distinguished in S if: for any three elements $a\ S\ b\ S\ c$ linearly ordered by S, if $a\ R\ c$ then $a\ R\ b$

REMARKS 1.2.2. a) We shall see later on that this finitary algebraic notion on the discrete structure R hides in fact some infinitary topological notion on the continuous structure $[R]$. But it has a number of non trivial properties which can be very pleasantly stated and proved in a purely algebraic language.

b) It is not difficult to construct distinguished subrelations of a given tree relation S. For example given any subet F of E, the (tree) relation $R := S \cap (F \times E)$ is obviously distinguished in S. The main aim of this section is to describe more elaborate constructions of distinguished subrelations.

c) In all applications of this notion both relations R and S will be tree relations. However it will be more convenient for the moment to consider arbitrary relations R. Notice that the distinction of R in S induces some reflexivity on R: For any $a \in E$, if $a\ R\ b$ for some b then $a\ S\ a\ S\ b$ hence necessarily $a\ R\ a$. We also have the following:

PROPOSITION 1.2.3. *Let R and S be two relations on a set E. Suppose that S is a tree relation and R is a partial ordering with a least element.*

If R is distinguished in S then R is a tree relation.

PROOF. Since $R \subset S$, the set of all R-predecessors of any element $c \in E$ is finite. Moreover if a and b are two R-predecessors of c then they are also S-predecessors of c, hence S-comparable, say $a\ S\ b\ S\ c$; then by distinction, $a\ R\ b$ and a and b are R-comparable. □

We now give a construction of explicit distinguished trees.

1.2.4. Lexicographical ordering. If $<$ is any strict linear ordering on the set F we define on the set $\mathrm{Seq}(F)$ the relation $<_{\mathrm{lex}}$ by:

$$s <_{\mathrm{lex}} t \iff \begin{cases} s \prec t \\ \text{or} \\ \exists\, k < |s|\,,\ \forall j < k,\ s(j) = t(j) \text{ and } s(k) < t(k) \end{cases}$$

It is clear that $<_{\mathrm{lex}}$ is a strict linear ordering on $\mathrm{Seq}(F)$, and we shall as usual denote by $\leq_{\mathrm{lex}}$ the corresponding non strict total ordering.

If $<$ is a well ordering on F then $<_{\mathrm{lex}}$ is a well ordering on $\mathrm{Seq}(F)$ too.

PROPOSITION 1.2.5. *Let S be a tree relation on a set E. Let $<$ be a strict linear ordering on a set F and $<_{\mathrm{lex}}$ be the corresponding linear ordering on* $\mathrm{Seq}(F)$. *If the mapping $\rho : E \longrightarrow \mathrm{Seq}(F)$ satisfies :*

$$a\ S\ b \implies \rho(a) \leq_{\mathrm{lex}} \rho(b)$$

then the relation S_ρ on E defined by:

$$a\ S_\rho\ b \iff (a\ S\ b \text{ and } \rho(a) \preceq \rho(b))$$

is a distinguished tree subrelation of S.

PROOF. We first prove that S_ρ is distinguished in S. Let a, b, c in E be such that $a\ S\ b\ S\ c$ and assume that $\rho(a) \preceq \rho(c)$; we have to show that $\rho(a) \preceq \rho(b)$. First notice that we have $\rho(a) \leq_{\mathrm{lex}} \rho(b) \leq_{\mathrm{lex}} \rho(c)$, and if $\rho(a) \not\preceq \rho(b)$ then there would exist $k < |\rho(a)|$ such that: $\rho(a)(j) = \rho(b)(j)$ for all $j < k$ and $\rho(a)(k) < \rho(b)(k)$, but since $\rho(a) \preceq \rho(c)$ then we would also have: $\rho(c)(j) = \rho(b)(j)$ for all $j < k$ and $\rho(c)(k) < \rho(b)(k)$, which contradicts $\rho(b) \leq_{\mathrm{lex}} \rho(c)$.

Finally since S_ρ is transitive and admits 0 as least element then by Proposition 1.2.3 S_ρ is a tree. □

The following result provides more distinguished tree relations and will be crucial for the representation theorem which we will prove in this chapter.

THEOREM 1.2.6. *Let $R \subset S$ be two tree relations on E. Then the smallest tree relation containing R and distinguished in S is the relation $\hat{R}$ defined by:*

$$a \hat{R} b \iff \begin{cases} \exists (b_i)_{0 \le i \le n}, \ \exists (c_i)_{1 \le i \le n} \text{ such that } b_0 = a, \quad b_n = b, \\ \forall i > 0, \quad b_{i-1} \, S \, b_i \, S \, c_i \quad \text{and} \quad b_{i-1} \, R \, c_i \end{cases}$$

PROOF. Let $\hat{R}$ be as above and define the relation $\tilde{R}$ by:

$$a \tilde{R} b \iff \exists c, \ a \, S \, b \, S \, c \text{ and } a \, R \, c$$

then

$$a \hat{R} b \iff \exists (b_i)_{0 \le i \le n} \text{ such that } \ a = b_0 \tilde{R} \dots b_{i-1} \tilde{R} b_i \dots \tilde{R} b_n$$

so $\hat{R}$ is the transitive closure of $\tilde{R}$ and clearly:

$$R \subset \tilde{R} \subset \hat{R} \subset S$$

□

LEMMA 1.2.7. *$\tilde{R}$ is the smallest relation containing R and distinguished in S.*

PROOF. We first prove that $\tilde{R}$ is distinguished in S. So suppose that $a \, S \, b \, S \, c$ and $a \tilde{R} c$. Fix d such that $a \, S \, c \, S \, d$ and $a \, R \, d$; then $a \, S \, b \, S \, d$ and so $a \tilde{R} b$.

Suppose now that $R \subset R' \subset S$ with R' distinguished in S. If $a \tilde{R} b$ fix c such that $a \, S \, b \, S \, c$ and $a \, R \, c$; since $R \subset R'$ it follows from the distinction of R' in S, that $a \, R' \, b$. This proves that $\tilde{R} \subset R'$.

LEMMA 1.2.8. *$\hat{R}$ is distinguished in S*

PROOF. Suppose that $a \, S \, b \, S \, c$ with $a \hat{R} c$, and fix a sequence $(a_i)_{0 \le i \le n}$ such that $a_0 = a \tilde{R} \dots a_{i-1} \tilde{R} a_i \dots \tilde{R} a_n = c$. Since $\tilde{R} \subset S$ there exists some $m \le n$ such that $a_{m-1} \ S \ b \ S \ a_m$, and since $\tilde{R}$ is distinguished in S then we also have $a_{m-1} \tilde{R} b$. Hence $a_0 = a \tilde{R} \dots a_{i-1} \tilde{R} a_i \dots a_{m-1} \tilde{R} b$ and so $a \hat{R} b$. □

To finish the proof of Theorem 1.2.6 observe that $\hat{R}$ is a partial ordering with a least element and is distinguished in S hence by Proposition 1.2.3 $\hat{R}$ is a tree relation.

Suppose now that $R \subset R' \subset S$ with R' distinguished tree subrelation of S. Then by Lemma 1.2.7 we have $\tilde{R} \subset R' \subset S$, and since $\hat{R}$ is the transitive closure of $\tilde{R}$ then $\hat{R} \subset R' \subset S$. This proves that $\hat{R}$ is the smallest tree relation containing R and distinguished in S. □

1.3. Canonical mapping of a distinguished subtree

In this section we prove the basic topological properties induced by the notion of distinction. Before we give formal statements and proofs we shall try to give the main intuition hidden behind.

Let E be the set $\omega^{<\omega}$ of all finite sequences of integers and suppose that R is a distinguished tree in Ext (the non strict extension relation). Interpret "$s' \, R \, s$" as "*s' is a good position when viewed from s*". Then for any $s \in E$ the R-branch of s (see 1.1.3) is the sequence of all good positions from the point of view of s, ordered by their lengths. Consider now two elements $s \prec t$ in $\omega^{<\omega}$;

let $\bar{s} := (\emptyset, s^{(1)}, s^{(2)}, \ldots, s^{(n-1)}, s)$ and $\bar{t} := (\emptyset, t^{(1)}, t^{(2)}, \ldots, t^{(m-1)}, t)$ denote their respective R-branches:

– if s is a good position from the point of view of t, then in this case the situation is trivial since s is one of the $t^{(k)}$, and because R is a tree relation then in fact $\bar{s} \preceq \bar{t}$.

– if not, then the comparison of $\bar{s}$ and $\bar{t}$ is unclear for an arbitrary tree relation $R \subset Ext$, and the role of distinction is precisely to bring some good behavior in this case by imposing that: " *If $s' \prec s$ is a good position when viewed from t, then necessarily s' was already a good position when viewed from s* ". Again because R is a tree relation then in fact there exist j and k such that $s^{(j)} \prec s \prec t^{(k)}$ so

$$\bar{t} = (\emptyset, s^{(1)}, s^{(2)}, \ldots, s^{(j)})^\frown(t^{(k)}, t^{(k+1)}, \ldots, t^{(m-1)}, t)$$

Thus at level t only a initial segment of what appeared to be good positions at level s are still viewed as good, in addition to some positions which are then longer than s. Notice however that applying the same observation again, one easily checks that the good positions $s^{(i)}$ with $i > j$ which disappeared at level t will never reappear at any further level $t' \succ t$.

The careful reader has surely recognized through these observations the starting up of a convergence procedure hence the definition of some Baire one mapping.

1.3.1. Canonical mapping of a subtree. If R is a subtree of S then any infinite R-branch is also an S-chain, hence is contained in a unique infinite S-branch. This defines a mapping

$$\pi : [R] \longrightarrow [S]$$

that we will call the <u>canonical mapping</u> of the subtree R of S.

Notice that if $\alpha \in [R] \subset E^\omega$ then $\pi(\alpha) \in [S] \subset E^\omega$ and in fact α is an infinite subsequence of $\pi(\alpha)$; more precisely $\alpha(k)$ is the unique coordinate of $\pi(\alpha)$ of R-height k. In particular the canonical mapping is *continuous*; but in general it is neither onto nor one-to-one.

If the the canonical mapping is onto we shall then say that S is an <u>expansion</u> of R.

REMARK 1.3.2. We shall prove soon that if R is distinguished in S then the canonical mapping $\pi : [R] \longrightarrow [S]$ is automatically one-to-one. But we emphasize that in general π need not be onto, so with our terminology S is not necessarily an expansion of R. Still in most applications we will need π to be a bijection; but this will be ensured by specific arguments.

THEOREM 1.3.3. *If R is a distinguished subtree of S and $\pi : [R] \to [S]$ is the corresponding canonical mapping then:*

a) π is one-to-one.

b) The range space $\pi([R])$ is a $\mathbf{\Pi}^0_3$ subset of $[S]$.

c) The image by π of any $\mathbf{\Sigma}^0_1$ subset of $[R]$ is a relatively $\mathbf{\Sigma}^0_2$ subset of $\pi([R])$, hence the inverse mapping $\pi^{-1} : \pi([R]) \to [R]$ is of Baire class one.

The proof of Theorem 1.3.3 follows essentially from the remarks made at the beginning of this section. But we shall extract from the arguments a general notion, that we present first, which will be a leading vector for other further proofs.

1.3.4. Distinguished R-chain of an S-chain. Let R be an arbitrary tree subrelation of S. If B is an S-chain then one easily checks that the set

$$A = \{a \in B \ : \ \forall b \in B \ (a \ S \ b \Rightarrow a \ R \ b)\}$$

is an R-chain, that we shall call the distinguished R-chain of B.

a) Notice that if $0 \in B$ then automatically $0 \in A$, but even if B is infinite the set A might be finite.

b) If R is a *distinguished* subtree of S, it is not difficult to check in this case that the R-branch of any element is precisely the distinguished R-chain of its S-branch. We shall now prove an infinitary version of this last observation.

PROPOSITION 1.3.5. *Let R be a distinguished subtree of S. If A is the distinguished R-chain of the S-chain B then :*

a) A is R-hereditary in B.

b) A contains any R-chain which is S-cofinal in B.

PROOF. a) Suppose that $a \in A$ and $a' \ R \ a$. Towards proving that $a' \in A$ consider $b \in B$ such that $a' \ S \ b$; then a, a', b are S-comparable and we distinguish two cases:

– If $a' \ S \ b \ S \ a$ then by distinction of R we have $a' \ R \ b$.

– If $a' \ S \ a \ S \ b$ then since $a \in A$ we have $a \ R \ b$ hence by transitivity $a' \ R \ b$.

This proves that $a' \in A$.

b) Let $A' \subset B$ be any R-chain which is S-cofinal in B and fix an element a in A'. For any $b \in B$ such that $a \ S \ b$ there exists $c \in A'$ such that $b \ S \ c$; since A' is an R-chain then a and c are R-comparable and since $a \ S \ c$ then we necessarily have $a \ R \ c$ hence by distinction $a \ R \ b$. This proves that $a \in A$; hence $A' \subset A$. □

COROLLARY 1.3.6. *Let R be a distinguished subtree of S. If B is an infinite S-branch then the distinguished R-chain A of B is an R-branch. Moreover:*

(i) either all R-branches contained in B, in particular A, are finite.

(ii) or A is the unique infinite R-branch contained in B.

PROOF. That A is an R-branch follows from Proposition 1.3.5 a.

Suppose that *(i)* does not hold and let A' be any infinite R-branch contained in B. Since A' is infinite it is S-cofinal in B, so by Proposition 1.3.5 b) we have $A' \subset A$; but since A' is a maximal R-chain then $A' = A$ which proves *(ii)*. □

Proof of Theorem 1.3.3: a) If $\alpha \in [R]$ enumerates some infinite R-branch A then $\beta = \pi(\alpha) \in [S]$ enumerates the unique infinite S-branch $B \supset A$, and by Corollary 1.3.6 A is the distinguished R-chain of B, hence α is uniquely determined by β.

b) Also by Corollary 1.3.6 for $\beta \in [S]$ we have:

$$\beta \in \pi([R]) \iff \forall k, \ \exists m, \Big(h_R(\beta(m)) = k \text{ and } \forall n \geq m, \ \beta(m) \ R \ \beta(n)\Big)$$

which shows that $\pi([R])$ is a $\mathbf{\Pi}^0_3$ subset of $[S]$.

c) Fix $a \in E$ and set $m = h_S(a)$ then again by Corollary 1.3.6 for $\beta \in \pi([R])$ we have:

$$\beta \in \pi(N_a^R) \iff \beta(m) = a \text{ and } \forall n \geq m, \ \beta(m) \ R \ \beta(n)$$

which shows that the image by π of any basic open subset of $[R]$ is a relatively closed subset of $\pi([R])$. Hence the image by π of any open (closed) subset of $[R]$ is a relatively $\mathbf{\Sigma}^0_2$ ($\mathbf{\Pi}^0_2$) subset of $\pi([R])$. □

DEFINITION 1.3.7. Let ξ be a countable ordinal.

a) We shall say that a family $(R^{(\eta)})_{\eta\leq\xi}$ of tree relations is a resolution family if:

1) $R^{(\eta+1)}$ is a distinguished subtree of $R^{(\eta)}$ for all $\eta<\xi$

2) $R^{(\lambda)}=\bigcap_{\eta<\lambda}R^{(\eta)}$ for all limit $\lambda\leq\xi$.

Let $R\subset S$ be two tree relations.

b) We shall say that R is ξ-distinguished in S if there exists a resolution family $(R^{(\eta)})_{\eta\leq\xi}$ with $R^{(0)}=S$ and $R^{(\xi)}=R$.

c) We shall say that S is a ξ-distinguished expansion of R if S is an expansion of R and R is ξ-distinguished in S.

We now prove a generalization of Theorem 1.3.3.

THEOREM 1.3.8. *Let $\pi:[R]\to[S]$ be the canonical mapping of the subtree R of S. If R is ξ-distinguished in S then:*

a) π is one-to-one.

b) The range space $\pi([R])$ is a $\mathbf{\Pi}^0_{\xi+2}$ subset of $[S]$.

c) The image by π of any $\mathbf{\Sigma}^0_1$ subset of $[R]$ is a relatively $\mathbf{\Sigma}^0_{\xi+1}$ subset of $\pi([R])$

PROOF. We fix a resolution family $(R^{(\eta)})_{\eta\leq\xi}$ for R in S, and denote for all $\eta\leq\xi$, by $\pi^{(\eta)}$ the canonical mapping from $[R^{(\eta)}]$ to $[S]$; in particular $\pi=\pi^{(\xi)}$. We shall prove the whole statement " a$\wedge$ b $\wedge$c)" by induction on ξ.

If $\xi=0$ then $R=S$ and π is the identity mapping and the statement is trivially true. Suppose it is true for all $\eta<\xi$, we distinguish the two cases:

Case 1: $\xi=\eta+1$ is a successor ordinal

Let σ denote the canonical mapping from $[R^{(\xi)}]$ to $[R^{(\eta)}]$ so that $\pi^{(\xi)}=\pi^{(\eta)}\circ\sigma$. Notice that $R^{(\xi)}$ is distinguished in $R^{(\eta)}$ and $R^{(\eta)}$ is η-distinguished in S, then applying Theorem 1.3.3 and the induction hypothesis we have:

a) σ and $\pi^{(\eta)}$ are one-to-one hence $\pi^{(\xi)}$ is also one-to-one.

b) By Theorem 1.3.3 *b)*, $\sigma([R])$ is $\mathbf{\Pi}^0_3$; it follows then from *c)* of the induction hypothesis that $\pi^{(\xi)}([R])=\pi^{(\eta)}(\sigma([R]))$ is a relatively $\mathbf{\Pi}^0_{\eta+3}$ subset of $\pi^{(\eta)}([R^{(\eta)}])$ which by *b)* is itself $\mathbf{\Pi}^0_{\eta+2}$, hence $\pi^{(\xi)}([R])$ is $\mathbf{\Pi}^0_{\eta+3}=\mathbf{\Pi}^0_{\xi+2}$.

c) As above if U is a $\mathbf{\Sigma}^0_1$ subset of $[R]$ then by Theorem 1.3.3 *c)* there exist a $\mathbf{\Sigma}^0_2$ set V such that $\sigma(U)=V\cap\sigma([R])$ and a $\mathbf{\Sigma}^0_{\eta+2}$ set W such that $\pi^{(\eta)}(V)=W\cap\pi^{(\eta)}([R^{(\eta)}])$; then $\pi^{(\xi)}(U)=W\cap\pi^{(\eta)}(\sigma([R^{(\eta)}]))\cap\pi^{(\eta)}(\sigma([R]))=W\cap\pi^{(\xi)}([R])$ is a relatively $\mathbf{\Sigma}^0_{\eta+2}=\mathbf{\Sigma}^0_{\xi+1}$ subset of $\pi^{(\xi)}([R])$.

Case 2: ξ is a limit ordinal

a) Fix some infinite S-branch B and let A' and A'' be two infinite $R^{(\xi)}$-branches contained in B. For any $\eta<\xi$, A' and A'' are two infinite $R^{(\eta)}$-chains hence by the induction hypothesis they are contained in a unique $R^{(\eta)}$-branch, and in particular their union $A=A'\cup A''$ is also an $R^{(\eta)}$-chain. Since this holds for all $\eta<\xi$ then A is also a $R^{(\xi)}$-chain, and by maximality $A'=A''=A$. This proves that π is one-to-one.

For all $\eta \le \xi$ and all $a \in E$ let $M_a^{(\eta)} \subset [S]$ denote the image by $\pi^{(\eta)}$ of the basic open set $N_a^{R^{(\eta)}}$ of $[R^{(\eta)}]$. By the induction hypothesis for all $\eta < \xi$, $M_a^{(\eta)}$ is a $\mathbf{\Pi}^0_{\eta+1}$ subset of $[S]$.

b) For $\beta \in [S]$ we have:

$$\beta \in \pi([R]) \iff \forall k,\ \exists a,\ h_{R^{(\xi)}}(a) = k \text{ and } \forall \eta < \xi,\ \beta \in M_a^{(\eta)}$$

which shows that $\pi([R])$ is a $\mathbf{\Pi}^0_{\xi+2}$ subset of $[S]$.

c) Fix $a \in E$; then for $\beta \in \pi([R])$ we have:

$$\beta \in M_a^{(\xi)} \iff \forall \eta < \xi\,,\ \beta \in M_a^{(\eta)}$$

which shows that $M_a^{(\xi)}$ is a relatively $\mathbf{\Pi}^0_\xi$ subset of $\pi([R])$. It follows then that the image by π of any open subset of $[R]$ is a relatively $\mathbf{\Sigma}^0_{\xi+1}$ subset of $\pi([R])$. □

1.4. Uniformly distinguished subtree

If ξ is finite then clause c) of Theorem 1.3.8 can be reformulated as in Theorem 1.3.3 by saying that the inverse mapping $\pi^{-1} : \pi([R]) \to [R]$ is of Baire class ξ. But if ξ is infinite this is no more the case (we recall that a mapping $\varphi : X \to Y$ is of class ξ iff the inverse image by φ of any open subset of Y is a $\mathbf{\Sigma}^0_{1+\xi}$ subset of X). Hence for ξ infinite the partial mapping π^{-1} is a priori of Baire class $\xi + 1$. In particular for λ a limit ordinal, Theorem 1.3.8 does not provide mappings π^{-1} of Baire class λ. For this one needs to strengthen the notion of distinction in the limit case.

Before we introduce this stronger notion, observe that if λ is limit then for all $b \in E$ the set $\{a :\ a\, R^{(\lambda)}\, b\}$ is just the intersection of the monotone family of finite sets $\big(\{a :\ a\, R^{(\eta)}\, b\}\big)_{\eta<\lambda}$; so there exists some $\eta < \lambda$ such that $\{a :\ a\, R^{(\lambda)}\, b\} = \{a : a\, R^{(\eta)}\, b\}$. Hence the following "left-uniformity" property:

$$\forall b\,,\ \exists \eta < \lambda\,,\ \forall a\,,\ (a\, R^{(\eta)}\, b \implies a\, R\, b)$$

always holds; but the "right" analog (obtained by swaping the roles of a and b) does not hold in general. However this "right-uniformity" property has a very interesting consequence :

PROPOSITION 1.4.1. *Let λ be a limit ordinal. If R is a λ-distinguished tree relation in S and admits a resolution family $(R^{(\eta)})_{\eta\le\lambda}$ satisfying:*

$$\forall a\,,\ \exists \eta < \lambda\,,\ \forall b\,,\ (a\, R^{(\eta)}\, b \implies a\, R\, b)$$

then the canonical mapping $\pi : [R] \to [S]$ is of Baire class λ.

PROOF. Let $a \in E$ and fix $\eta < \lambda$ such that $(a\, R^{(\eta)}\, b \implies a\, R\, b)$ then (following the notations of the proof of Theorem 1.3.8) $M_a^{(\lambda)} = M_a^{(\eta)}$ which by Theorem 1.3.8 is a $\mathbf{\Pi}^0_{\eta+1}$ subset of $\pi([R])$. It follows then that the image by π of any open subset of $[R]$ is a relatively $\mathbf{\Sigma}^0_\lambda$ subset of $\pi([R])$. □

In fact, for several applications, we will need the following stronger notion of uniformity:

DEFINITION 1.4.2. Let λ be an infinite limit ordinal.

a) We shall say that the resolution family $(R^{(\eta)})_{\eta\le\lambda}$ is uniform if:

$$\forall n\,,\ \exists \eta < \lambda, \forall a\,,\ \forall b \ \text{ such that } \min(h_R(a)\,,\ h_R(b)) \le n\,,\ (a\, R^{(\eta)}\, b \implies a\, R\, b)$$

Let $R \subset S$ be two tree relations.

b) We shall say that R is uniformly λ-distinguished in S if there exists a uniform resolution family $(R^{(\eta)})_{\eta \leq \lambda}$ with $R^{(0)} = S$ and $R^{(\lambda)} = R$.

c) We shall say that S is a uniform λ-distinguished expansion of R if S is an expansion of R and R is uniformly λ-distinguished in S.

We emphasize that a uniform resolution family $(R^{(\eta)})_{\eta \leq \lambda}$ is not required to satisfy any uniformity property at limit levels $\lambda' < \lambda$, but only at the last level λ.

We now prove a result which will constitute a crucial ingredient in the proof of the main result of next section. Notice that if $R \subset \tilde{R} \subset S$ are tree relations with R distinguished in $\tilde{R}$ and $\tilde{R}$ λ-distinguished in S then R is $(\lambda+1)$-distinguished in S. However:

THEOREM 1.4.3. *Let $R \subset \tilde{R} \subset S$ be three tree relations and λ be a limit ordinal. If R is distinguished in $\tilde{R}$ and $\tilde{R}$ is uniformly λ-distinguished in S then R is λ-distinguished in S*

PROOF. Fix a uniform resolution family $(S^{(\xi)})_{\xi \leq \lambda}$ of $\tilde{R}$ in S, and define inductively a family $(R^{(\xi)})_{\xi \leq \lambda}$ by:

$$R^{(\xi)} = \begin{cases} S^{(0)} & \text{if } \xi = 0 \\ \text{the smallest tree relation containing} & \\ R \text{ and distinguished in } S^{(\eta)} & \text{if } \xi = \eta + 1 \\ \bigcap_{\eta < \xi} R^{(\eta)} & \text{if } \xi > 0 \text{ is limit} \end{cases}$$

We recall that by Theorem 1.2.6 for all $\eta < \lambda$ we have:

$$a\, R^{(\eta+1)}\, b \iff \begin{cases} \exists\, (b_i)_{0 \leq i \leq n}\,,\ \exists\, (c_i)_{1 \leq i \leq n} \text{ such that } b_0 = a, \quad b_n = b, \\ \forall i > 0\,, \quad b_{i-1}\, S^{(\eta)}\, b_i\, S^{(\eta)}\, c_i \quad \text{and} \quad b_{i-1}\, R\, c_i \end{cases}$$

moreover it is clear that one can suppose the sequence $(b_i)_{0 \leq i \leq n}$ to be one-to-one. In particular for $a \neq b$, applying the previous equivalence inductively we have:

$$a\, R^{(\eta+1)}\, b \iff \exists\, b' \neq b\,, \exists\, c' \text{ such that } a\, R^{(\eta+1)}\, b'\, S^{(\eta)}\, b \text{ and } b'\, R\, c'$$

(take $b' = b_{n-1}$ and $c' = c_n$).

CLAIM 1. *For all $\xi \leq \lambda$, $R^{(\xi)} \subset S^{(\xi)}$ and $R^{(\xi)}$ is distinguished in $S^{(\xi)}$.*

Proof: By induction on ξ: for $\xi = 0$ this is obvious. Suppose that $R^{(\eta)}$ is distinguished in $S^{(\eta)}$ for all $\eta < \xi$.

– If ξ is limit, then clearly $R^{(\xi)} = \bigcap_{\eta < \xi} R^{(\eta)}$ is distinguished in $S^{(\xi)} = \bigcap_{\eta < \xi} S^{(\eta)}$.

– If $\xi = \eta + 1$, notice first that since $(S^{(\zeta)})_{\zeta \leq \lambda}$ is a resolution family of $\tilde{R}$ then $R \subset \tilde{R} \subset S^{(\eta+1)}$ and $S^{(\eta+1)}$ is distinguished in $S^{(\eta)}$. Hence $R^{(\eta+1)} \subset S^{(\eta+1)} \subset S^{(\eta)}$ and since $R^{(\eta+1)}$ is distinguished in $S^{(\eta)}$ then a fortiori $R^{(\eta+1)}$ is distinguished in $S^{(\eta+1)}$.

◇

CLAIM 2. *$(R^{(\xi)})_{\xi \leq \lambda}$ is a resolution family of $R^{(\lambda)}$ in S.*

Proof: By definition $R^{(0)} = S^{(0)} = S$ and $R^{(\xi)} = \bigcap_{\eta<\xi} R^{(\eta)}$ if $\xi > 0$ is limit.

For the successor case notice that by Claim 1 we have $R \subset R^{(\eta)} \subset S^{(\eta)}$ and $R^{(\eta)}$ is distinguished in $S^{(\eta)}$, hence $R^{(\eta+1)} \subset R^{(\eta)}$ and a fortiori $R^{(\eta+1)}$ is distinguished in $R^{(\eta)}$. ◇

CLAIM 3. $R^{(\lambda)} = R$

Proof: The inclusion $R \subset R^{(\lambda)}$ is obvious. For the converse inclusion we fix a and shall prove the property:

$$\mathrm{P}(b): \qquad (a\, R^{(\lambda)}\, b \implies a\, R\, b)$$

by induction on S: Assuming that P holds for all S-predecessors of b we shall prove $\mathrm{P}(b)$.

So suppose that $a\ R^{(\lambda)}\ b$ with $a \neq b$ (otherwise $a\ R\ b$ trivially). Since $\tilde{R}$ is uniformly λ-distinguished in S we can fix η such that:

$$\forall c \quad \big(b\, R^{(\eta)}\, c \implies b\, R\, c\big)$$

From the observations above, for all $\xi < \lambda$ we can also fix $b'_\xi \neq a$ and c'_ξ such that: $a\ R^{(\xi+1)}\ b'_\xi\ S^{(\xi)}\ b$ and $b'_\xi\ R\ c'_\xi$. Then the monotone family of finite sets $\big(\{a' : a'\ S^{(\xi)}\ b\}\big)_{\xi<\lambda}$ is eventually constant then $b'_\xi = b'$ is also constant for all $\xi \in \Lambda$ a cofinal subset of λ. But since $R^{(\lambda)} = \bigcap_{\xi\in\Lambda} R^{(\xi+1)}$ then $a\ R^{(\lambda)}\ b'$ and it follows then from the induction hypothesis that $a\ R\ b'$. Similarly since $\tilde{R} = S^{(\lambda)} = \bigcap_{\xi\in\Lambda} S^{(\xi)}$ then $b'\,\tilde{R}\,b$. Fix now $\zeta \in \Lambda$ with $\zeta > \eta$ and set $c = c'_\zeta$; then $a\,\tilde{R}\,b'\,\tilde{R}\,b$ and $a\ R\ c$, hence by the distinction of R in $\tilde{R}$ we also have $a\ R\ b$. This finishes the proof of $\mathrm{P}(b)$, consequently of Claim 3. ◇

Theorem 1.4.3 clearly follows from Claim 2 and Claim 3 . □

1.5. Tree products

In this section we shall describe a general construction which provides uniform resolution families. This construction will be relative to some *fixed* tree relation S and *all trees considered in this section will implicitly be supposed to be subtrees of* S

1.5.1. Index mapping of a subtree: The *index mapping* of a subtree R of S is the unique mapping $\varphi_R : E \to E$ defined by:

$$\begin{cases} \varphi_R(0) = 0 \\ \\ \varphi_R(a)\ S\ a \quad \text{and} \quad h_S(\varphi_R(a)) = h_R(a) - 1\,, \quad \text{if } a \neq 0 \end{cases}$$

We first check that this definition is consistent: Notice that for any $m \leq h_S(a)$ there exists a unique a' satisfying: ($a'\ S\ a$ and $h_S(a') = m$); on the other hand since $R \subset S$ then $h_R(a) \leq h_S(a)$, and so $\varphi_R(a)$ is well defined.

Notice that if $R = S$ then in this particular case $\varphi_S(a)$ is just the S-predecessor of a, so the mapping φ_S is the S-predecessor function (see 1.1.3). In the general case $\varphi_S(a)$ encodes the R-chain of a by the S-chain of some S-predecessor of a, namely $\varphi_R(a)$. The role of the "-1" in the formula above is to ensure that $\varphi_R(a)$ is a *strict* predecessor of a. This last detail is fundamental for our purposes as it will become clear later, although it will be the source of some unpleasant algebraic complications.

LEMMA 1.5.2. *If* $a\,R\,b$ *then* $\varphi_R(a)\ S\ \varphi_R(b)$

PROOF. If $a = 0$ then $\varphi_R(a) = 0$ and the lemma is obvious; so we may suppose that $a \neq 0$. From the definition of the index mapping we have $\varphi_R(a)\ S\ a\ S\ b$ and $\varphi_R(b)\ S\ b$, hence $\varphi_R(a)$ and $\varphi_R(b)$ are S-comparable; and since $a\,R\,b$ we also have $h_S(\varphi_R(a)) = h_R(a) - 1 \leq h_R(b) - 1 = h_S(\varphi_R(b))$ hence $\varphi_R(a)\ S\ \varphi_R(b)$. □

LEMMA 1.5.3. *For any* $a \neq 0$*, if* A *is the* R*-branch of* a *and* B *is the* S*-branch of* $\varphi_R(a)$ *then* φ_R *induces a bijection from* $A \setminus \{0\}$ *onto* B.

PROOF. Let $b \in \varphi_R(A\setminus\{0\})$ and fix $a' \in A\setminus\{0\}$ such that $b = \varphi_R(a')$. It follows from the definition of φ_R that $h_R(a') = h_S(b) + 1$ hence a' is uniquely determined by b. This shows that φ_R is one-to-one on $A \setminus \{0\}$. Moreover by Lemma 1.5.2, $\varphi_R(A \setminus \{0\}) \subset B$ and since

$$\mathrm{card}(\varphi_R(A \setminus \{0\})) = \mathrm{card}(A) - 1 = h_R(a) - 1 = h_S(\varphi_R(a) = \mathrm{card}(B)$$

then $\varphi_R(A \setminus \{0\}) = B$. □

1.5.4. The $\otimes$ operation: If R_0 and R_1 are two subtrees of S we define the relation $R_0 \otimes R_1$ of S by:

$$a\,R_0 \otimes R_1\,b \iff (\ a\,R_0\,b \ \text{ and } \ \varphi_{R_0}(a)\ R_1\ \varphi_{R_0}(b)\ )$$

LEMMA 1.5.5. $R_0 \otimes R_1$ *is a tree relation.*

PROOF. It is clear that $R_0 \otimes R_1$ is a partial ordering on E, and that $R_0 \otimes R_1 \subset R_0$ hence the set of all $R_0 \otimes R_1$-predecessors of any element is finite.

Suppose that a and b are two $R_0 \otimes R_1$-predecessors of some element c; we have to check that a and b are $R_0 \otimes R_1$-comparable. It follows from the definition of $R_0 \otimes R_1$ that a and b are R_0-predecessors of c hence R_0-comparable, say $a\ R_0\ b\ R_0\ c$; and similarly that $\varphi_{R_0}(a)$ and $\varphi_{R_0}(b)$, which are R_1-predecessors of $\varphi_{R_0}(c)$, are R_1-comparable. Moreover by Lemma 1.5.2 we have $\varphi_{R_0}(a)\ S\ \varphi_{R_0}(b)$, and since $R_1 \subset S$ then necessarily $\varphi_{R_0}(a)\ R_1\ \varphi_{R_0}(b)$, and so $a\,R_0 \otimes R_1\,b$. □

LEMMA 1.5.6. $R_0 \otimes S = R_0$

PROOF. By Lemma 1.5.2 we have:

$$a\ R_0 \otimes S\ b \iff \Big(a\ R_0\ b \ \text{ and } \ \varphi_{R_0}(a)\ S\ \varphi_{R_0}(b)\Big) \iff a\ R_0\ b$$

□

LEMMA 1.5.7. *If* R' *is* ξ*-distinguished in* R *then* $R_0 \otimes R'$ *is* ξ*-distinguished in* $R_0 \otimes R$.

More precisely if $(R^{(\eta)})_{\eta\leq\xi}$ *is a resolution family for* R' *in* R *then* $(R_0 \otimes R^{(\eta)})_{\eta\leq\xi}$ *is a resolution family for* $R_0 \otimes R'$ *in* $R_0 \otimes R$.

PROOF. First notice that for any family $(R_i)_{i\in I}$ of trees, $R_0 \otimes \bigcap_{i\in I} R_i = \bigcap_{i\in I} R_0 \otimes R_i$. In particular if $R' \subset R$ then $R_0 \otimes R' \subset R_0 \otimes R$, and it follows directly from definitions that if R' is distinguished in R then $R_0 \otimes R'$ is distinguished in $R_0 \otimes R$.

Hence for all η, $R_0 \otimes R^{(\eta+1)}$ is distinguished in $R_0 \otimes R^{(\eta)}$, and by the formula above for limit λ we have: $R_0 \otimes R^{(\lambda)} = R_0 \otimes \bigcap_{\eta<\lambda} R^{(\eta)} = \bigcap_{\eta<\lambda} R_0 \otimes R^{(\eta)}$ □

We shall now extend the operation $\otimes$ to finite families of trees. Since this operation is not associative this needs some explanations. Notice that the fact that the operation $\otimes$ is not associative is due to the presence of the "-1" in the formula "$h_S(\varphi_R(a)) = h_R(a) - 1$" defining the mapping φ_R. In fact by deleting this "-1" one would get an associative operation for which S happens to be a two-sided unit, and which has many nice algebraic properties such as $\varphi_{R_0 \otimes R_1} = \varphi_{R_0} \circ \varphi_{R_1}$ which do not hold with our definition. But this alternative $\otimes$ operation (without "-1") cannot be extended to define a $\otimes$ product for *infinite* families of trees because in this case the process would become ill-founded.

1.5.8. Finite products: We define inductively on n, the product $\bigotimes_{i=0}^{n} R_i$ of any finite sequence $(R_i)_{0 \leq i \leq n}$ of subtrees of S by $\bigotimes_{i=0}^{0} R_i = R_0$ if $n = 0$, and for $n > 0$ by:

$$\bigotimes_{i=0}^{n} R_i = R_0 \otimes (\bigotimes_{i=0}^{n-1} R_{i+1})$$

It will also be convenient for the sequel to introduce the following alternative notation:

$$\otimes (R_0, R_1, \ldots, R_n) := \bigotimes_{i=0}^{n} R_i$$

In particular $R_0 \otimes R_1 = \otimes (R_0, R_1) = \bigotimes_{i=0}^{1} R_i$.

To get a more explicit description of $\bigotimes_{i=0}^{n} R_i$ define inductively:

$$\psi_m(a) = \begin{cases} a & \text{if } m = 0 \\ (\varphi_{R_{m-1}} \circ \psi_{m-1})(a) & \text{if } 0 < m \leq n \end{cases}$$

so more explicitly $\psi_m = \varphi_{R_{m-1}} \circ \cdots \circ \varphi_{R_1} \circ \varphi_{R_0}$ if $m > 0$. It follows then immediately from the definitions above that:

$$a \, (\bigotimes_{i=0}^{n} R_i) \, b \iff \forall \, k \leq n, \;\; \psi_k(a) \, R_k \, \psi_k(b)$$

Notice also that if $\psi_m(a) = 0$ then $\psi_k(a) = 0$ for all $k \geq m$.

The following two results can be derived by an obvious induction from Lemma 1.5.6 and Lemma 1.5.7 :

LEMMA 1.5.9. $\otimes (R_0, \ldots, R_n, S) = \otimes (R_0, \ldots, R_n)$

LEMMA 1.5.10. *If R' is ξ-distinguished in R then $\tilde{R}' := \otimes (R_0, \ldots, R_{n-1}, R')$ is ξ-distinguished in $\tilde{R} := \otimes (R_0, \ldots, R_{n-1}, R)$. More precisely:*

If $(R^{(\eta)})_{\eta \leq \xi}$ is a resolution family for R' in R then $(\otimes (R_0, \ldots, R_{n-1}, R^{(\eta)}))_{\eta \leq \xi}$ is a resolution family for $\tilde{R}'$ in $\tilde{R}$.

1.5.11. Infinite products: Let $(R_i)_{i\in\omega}$ be an *infinite* sequence of subtrees of S. Then from the definitions we have $\bigotimes_{i=0}^{n+1} R_i \subset \bigotimes_{i=0}^{n} R_i$ for all $n \in \omega$; and we shall define:

$$\bigotimes_{i=0}^{\infty} R_i := \bigcap_{n\in\omega} \bigotimes_{i=0}^{n} R_i$$

One clearly has an explicit description of $\bigotimes_{i=0}^{\infty} R_i$ in terms of the ψ_n as in 1.5.8. Observe also that if we extend the *finite* sequence $(R_i)_{0\leq i\leq n}$ by setting $R_i = S$ for $i > n$, then it follows from Lemma 1.5.6 that $\bigotimes_{i=0}^{n} R_i = \bigotimes_{i=0}^{\infty} R_i$. Thus finite products are particular cases of infinite products.

THEOREM 1.5.12. *Let $h_{\tilde{R}}$ denote the height function of the infinite product $\tilde{R} := \bigotimes_{i=0}^{\infty} R_i$, and set for all n, $E_n = \{a \in E : \ h_{\tilde{R}}(a) \leq n\}$ then*

$$(E \times E_n) \cap \bigotimes_{i=0}^{n} R_i \ \subset \ (E_n \times E) \cap \bigotimes_{i=0}^{n} R_i \ \subset \ \bigotimes_{i=0}^{\infty} R_i$$

PROOF. For all n set: $\tilde{R}_n = \bigotimes_{i=0}^{n} R_i$ and $\psi_n = \varphi_{R_{n-1}} \circ \ldots \circ \varphi_{R_1} \circ \varphi_{R_0}$ (see 1.5.8 for basic properties of the ψ_n). In fact we shall give an explicit formula for $h_{\tilde{R}}$:

LEMMA 1.5.13. $h_{\tilde{R}}(a) = \min\{k : \ \psi_k(a)) = 0\}$

We first derive Theorem 1.5.12 from Lemma 1.5.13. For this observe that by the lemma we have: $E_n = \{a \in E : \ \psi_n(a) = 0\}$.

To prove the first inclusion of the Theorem, notice that if $a \ \tilde{R}_n \ b$ then $\psi_n(a) \ R_n \ \psi_n(b)$; and if moreover $\psi_n(b) = 0$ then necessarily $\psi_n(a) = 0$.

For the second inclusion notice that if $a \ \tilde{R}_n \ b$ then for all $k \leq n$, $\psi_k(a) \ R_k \ \psi_k(b)$; and if moreover $\psi_n(a) = 0$ then for all $k > n$ we also have: $0 = \psi_k(a) \ R_k \ \psi_k(b)$; hence $a \ \tilde{R} \ b$.

We now proceed to the proof of Lemma 1.5.13:

CLAIM 1. *If $a \neq 0$ then $h_{R_0 \otimes R_1}(a) = h_{R_1}(\varphi_{R_0}(a)) + 1$.*

Let A denote the $R_0 \otimes R_1$-branch of a and B denote the R_1-branch of $\varphi_{R_0}(a)$. It follows from the definition of $R_0 \otimes R_1$ that A is contained in the R_0-branch of a, B is contained in the S-branch of a and $\varphi_{R_0}(A) = B$. On the other hand by Lemma 1.5.3 φ_{R_0} is one-to-one from $A \setminus \{0\}$ onto B hence:

$$h_{R_0 \otimes R_1}(a) = \text{card}\,(A) = \text{card}\,(B) + 1 = h_{R_1}(\varphi_{R_0}(a)) + 1 \qquad \diamond$$

CLAIM 2. *For all $n > 0$, if $\psi_{n-1}(a) \neq 0$ then $h_{\tilde{R}_n}(a) = h_{R_n}(\psi_n(a)) + n$.*

By Claim 1 this is true for $n = 1$. So suppose that the result holds for $n-1$; set $\tilde{S} = \bigotimes_{i=0}^{n-1} R_{i+1}$ so that $\tilde{R}_n = R_0 \otimes \tilde{S}$. Since $\psi_{n-1}(a) \neq 0$ then $a \neq 0$, so by Claim 1 we have: $h_{\tilde{R}_n}(a) = h_{R_0 \otimes \tilde{S}}(a) = h_{\tilde{S}}(\varphi_{R_0}(a)) + 1$. Again since $\psi_{n-1}(a) \neq 0$ then

$b := \varphi_{R_{n-2}} \circ \varphi_{R_{n-3}} \circ \cdots \circ \varphi_{R_1}(\varphi_{R_0}(a)) \neq 0$ and $\varphi_{R_{n-1}}(b) = \psi_n(a)$; then applying the induction hypothesis to the family $(R_{i+1})_{0 \leq i \leq n-1}$ and to the element $\varphi_{R_0}(a)$ we have: $h_{\tilde{R}}(\varphi_{R_0}(a)) = h_{R_n}(\psi_n(a)) + n - 1$ hence $h_{\tilde{R}_n}(a) = h_{R_n}(\psi_n(a)) + n$. ◇

To finish the proof of Lemma 1.5.13, set $n := \min\{k : \psi_k(a) = 0\}$.

If $a' \, \tilde{R}_n \, a$ then $\psi_n(a') \; R_n \; \psi_n(a) = 0$ hence $\psi_n(a') = 0$ and a fortiori $\psi_k(a') = 0$ for all $k \geq n$. Hence $h_{\tilde{R}}(a) = h_{\tilde{R}_n}(a)$; and since $\psi_{n-1}(a) \neq 0$ then by Claim 2, $h_{\tilde{R}}(a) = h_{\tilde{R}_n}(a) = h_{R_n}(\psi_n(a)) + n = n$. □

THEOREM 1.5.14. *If for all i, R_i is ξ_i-distinguished in S for some $\xi_i > 0$, then $\tilde{R} := \bigotimes_{i=0}^{\infty} R_i$ is uniformly λ-distinguished in S, where $\lambda = \sum_{i \in \omega} \xi_i$ is the ordinal sum of $(\xi_i)_{i \in \omega}$.*

PROOF. Set $\eta_{-1} = 0$, and $\eta_n = \sum_{k=0}^{n} \xi_k$ if $n \geq 0$. We fix for all $n \geq 0$ a resolution family $(R_n^{(\xi)})_{\xi \leq \xi_n}$ for R_n in S, and define :

$$\begin{cases} \tilde{R}^{(\eta)} = \otimes(R_0, R_1, \ldots, R_{n-1}, R_n^{(\xi)}) & \text{if } \eta = \eta_{n-1} + \xi \text{ with } \xi < \xi_n \\ \tilde{R}^{(\lambda)} = \bigotimes_{i=0}^{\infty} R_i & \end{cases}$$

Since $R_{n+1}^{(0)} = S$ then $\tilde{R}^{(\eta_n)} = \otimes(R_0, \ldots, R_{n-1}, R_n, R_{n+1}^{(0)}) = \otimes(R_0, \ldots, R_{n-1}, R_n)$ and since $R_n = R_n^{(\xi_n)}$ then $\tilde{R}^{(\eta_n)} = \otimes(R_0, \ldots, R_{n-1}, R_n^{(\xi_n)})$. Then by Lemma 1.5.7, $(\tilde{R}^{(\eta)})_{\eta_n \leq \eta \leq \eta_{n+1}}$ (with *non strict* inequalities for indices) is a resolution family; moreover for the limit ordinal λ we also have $\tilde{R}^{(\lambda)} = \bigcap_{n \geq 0} \tilde{R}^{(\eta_n)} = \bigcap_{\eta < \lambda} \tilde{R}^{(\eta)}$. Hence $(\tilde{R}^{(\eta)})_{\eta \leq \lambda}$ is a resolution family for $\tilde{R}$ in S.

Finally it follows from Theorem 1.5.12 that for all $n \geq 0$ and all a and b in E if $\min(h_{\tilde{R}}(a), h_{\tilde{R}}(b)) \leq n$ and $a \, \tilde{R}^{(\eta_n)} b$ then $a \, \tilde{R} \, b$. Hence $\tilde{R}$ is uniformly λ-distinguished in S. □

We now present the second important property of infinite products. We recall that S is an expansion of R means that the canonical mapping $\pi : [R] \to [S]$ is onto.

THEOREM 1.5.15. *If for all i, S is an expansion of R_i then S is an expansion of $\bigotimes_{i=0}^{\infty} R_i$.*

PROOF. We follow the notations introduced in the proof of Theorem 1.5.12.

Fix an infinite S-branch A; we have to prove that A contains an infinite $\tilde{R}$-branch. We first construct an infinite sequence $(A_n)_{n \in \omega}$ satisfying for all n:

$$\begin{cases} A_n \text{ is an infinite } R_n\text{-branch} \\ A_0 \subset A \,, \quad \text{and} \quad A_{n+1} \subset \varphi_{R_n}(A_n) \end{cases}$$

For A_0 we take any infinite R_0-branch contained in A, which is possible since S is an expansion of R_0. Suppose that A_n is constructed; then by Lemma 1.5.3, $\varphi_{R_n}(A_n)$ is an S-branch, and again since S is an expansion of R_{n+1} we can find an infinite R_{n+1}-branch A_{n+1} contained in $\varphi_{R_n}(A_n)$, and the construction is finished.

We then construct for all $n \geq 0$ an element $a_n \in A_0$ such that:

$$\begin{cases} \psi_k(a_n) \in A_k \setminus \{0\} & \text{for all } k < n \\ \psi_n(a_n) = 0 & \end{cases}$$

For this fix some n, and notice that since $A_k \subset \varphi_{R_{n-1}}(A_{k-1})$ for all $k \leq n$, then starting from $b_0 = 0 \in A_n$ and applying Lemma 1.5.3 one can construct by a straightforward induction a finite sequence $(b_k)_{k \leq n}$ such that:

$$\begin{cases} b_k \in A_{n-k} \setminus \{0\} & \text{for all } k > 0 \\ b_k = \varphi_{R_{n-1}}(b_{k+1}) & \text{for all } k < n \end{cases}$$

Now if we set $a_n = b_n$ then $\psi_k(a_n) = b_{n-k}$ for all $k \leq n$ and we are done.

By construction $\tilde{A} = \{a_n \,;\, n \geq 0\} \subset A_0 \subset A$ is infinite, and we now show that it is an $\tilde{R}$-chain (in fact one can show that $\tilde{A}$ is an $\tilde{R}$-branch). Fix two integers m and n; since A_0 is an R_0-chain then a_m and a_n are R_0-comparable, so suppose that $a_m \; R_0 \; a_n$. Notice that similarly, by construction for all $k \geq 0$, $\psi_k(a_m)$ and $\psi_k(a_n)$ are R_k-comparable, and in fact:

CLAIM 3. $\psi_k(a_m) \; R_k \; \psi_k(a_n)$

We prove the claim by induction on k. This is obvious for $k = 0$; so suppose that $\psi_{k-1}(a_m) \; R_{k-1} \; \psi_{k-1}(a_n)$. Recall that $\psi_k = \varphi_{R_{k-1}} \circ \psi_{k-1}$ hence by Lemma 1.5.2 we also have $\psi_k(a_m) \; S \; \psi_k(a_n)$ and since $\psi_k(a_m)$ and $\psi_k(a_n)$ are R_k-comparable then necessarily $\psi_k(a_m) \; R_k \; \psi_k(a_n)$. ◇

It follows from the claim that $a_m \, \tilde{R} \, a_n$; hence $\tilde{A}$ is an $\tilde{R}$-chain and this finishes the proof of the theorem. □

1.6. Tree expansions and representations of Borel sets

The following definition, which is quite close to Definitions 1.3.7 and 1.4.2, will enable us to state simply the representation theorem of Borel sets.

DEFINITION 1.6.1. labeldef Let ξ be a countable ordinal, $R \subset S$ be two tree relations with $\pi : [R] \to [S]$ the canonical mapping, and $(B_i)_{i \in I}$ be a family of subsets of $[S]$.

a) We shall say that $\langle S, (B_i)_{i \in I} \rangle$ is a (uniform) ξ-distinguished expansion of $\langle R, (A_i)_{i \in I} \rangle$ if S is a (uniform) ξ-distinguished expansion of R and for all $i \in I$, and $A_i = \pi^{-1}(B_i) \subset [R]$.

b) For any class Γ, we shall say that the family $(B_i)_{i \in I}$ is a (uniform) ξ-distinguished expansion of Γ sets, if $\langle S, (B_i)_{i \in I} \rangle$ is a (uniform) ξ-distinguished expansion of some $\langle R, (A_i)_{i \in I} \rangle$ with $A_i \in \Gamma$ for all i.

THEOREM 1.6.2. *Any countable family of $\mathbf{\Pi}^0_\xi$ sets is a ξ-distinguished expansion of $\mathbf{\Delta}^0_1$ sets.*

Notice that by symmetry the similar statement for additive classes $\mathbf{\Sigma}^0_\xi$ also holds. We first prove the particular case $\xi = 1$:

LEMMA 1.6.3. *Any countable family of $\mathbf{\Pi}^0_1$ sets is a distinguished expansion of $\mathbf{\Delta}^0_1$ sets.*

PROOF. Let $(A_n)_{n \in \omega}$ be a countable family of $\mathbf{\Pi}^0_1$ sets in $[S]$. Fix for all n, a partition $(A_{(n,k)})_{k \in \omega}$ of $[S]$ with: $A_{(n,0)} = A_n$, and $A_{(n,k)}$ $\mathbf{\Delta}^0_1$ for $k > 0$; and define for all $s \in \mathrm{Seq}(\omega)$:

$$F(s) = \bigcap_{i < |s|} A_{(i, s(i))}$$

then for all n, $(F(s))_{s\in\omega^n}$ is a family of pairwise disjoint closed sets covering $[S] = F(\emptyset)$.

For all $a \in E$ we denote by N_a simply the basic open set N_a^S of $[S]$, and define:

$$\sigma(a) = \min\{s \in \omega^n : \quad N_a \cap F(s) \neq \emptyset\} \qquad \text{where } n = h_S(a)$$

and the minimum is relative to the lexicographical ordering on ω^n which, we recall, is induced by the lexicographical ordering $\leq_{\text{lex}}$ on $\text{Seq}(\omega)$ (see 1.2.4). Notice that if $a\, S\, b$ with $n = h_S(a) \leq h_S(b) = m$ then for any $t \in \omega^m$, $N_a \cap F(t_{|n}) \supset N_b \cap F(t)$; and it follows that $\sigma(a) \leq_{\text{lex}} \sigma(b)_{|n} \leq_{\text{lex}} \sigma(b)$.

Next we define, inductively on S, a mapping $\rho : E \to \text{Seq}(\omega)$ by:

$$\rho(a) = \begin{cases} \emptyset & \text{if } a = 0 \\ \sigma(a)_{|p+1} & \text{if } a \neq 0\,,\text{ where} \\ & \quad p = \max\{q : \exists\, a' \neq a,\ a'\ S\ a \text{ and } \rho(a') = \sigma(a)_{|q}\} \end{cases}$$

Again if $a\, S\, b$ with $\rho(a) = \sigma(a)_{|p+1}$ and $\rho(b) = \sigma(b)_{|q+1}$ then necessarily $p \leq q$, and since $\sigma(a) \leq_{\text{lex}} \sigma(b)$ then we also have $\rho(a) \leq_{\text{lex}} \rho(b)$. Finally if we define the relation R by:

$$a\, R\, b \iff a\, S\, b \ \text{ and } \ \rho(a) \leq_{\text{lex}} \rho(b)$$

then by Proposition 1.2.5 R is a distinguished subtree of S.

We fix now $\beta = (a_n)_{n\in\omega}$ in $[S]$, and let $\alpha \in E^{\leq\omega}$ be the enumeration of the distinguished R-chain of the S-branch enumerated by β (see Proposition 1.3.4). We recall that the sequence α is possibly finite, and denote by $|\alpha|$ its length (with $|\alpha| = \infty$ if α is infinite).

CLAIM . *If $\beta \subset F(t)$ for some $t \in \omega^p$ then $|\alpha| > p$ and $\rho(\alpha(p)) = t$.*

Proof: We prove the claim by induction on p: This is true for $p = 0$ since $\rho(0) = \emptyset$. Suppose it's true for p and notice that there exists a unique m_p such that $\alpha(p) = \beta(m_p)$.

If $\beta \in F(t^\frown\langle k\rangle)$ for some $k \in \omega$ then $\beta \notin F = \bigcup_{j<k} F(t^\frown\langle j\rangle)$, and since F is a closed subset of $[S]$ there exists a neighborhood of β which is disjoint from F. Define:

$$m_{p+1} = \min\{m > m_p : \quad \beta(m_p)\ S\ \beta(m) \text{ and } N_{\beta(m)} \cap F = \emptyset\}$$

and observe that for all $n \geq m_{p+1}$ we have:

(1) $N_{\beta(n)} \cap F(t^\frown\langle k\rangle) \neq \emptyset$ (since it contains β)

(2) $\forall j < k\,,\ \ N_{\beta(n)} \cap F(t^\frown\langle j\rangle) \subset N_{\beta(m_{p+1})} \cap F = \emptyset$

(3) $|\sigma(\beta(n))| = h_S(\beta(n)) = n \geq m_{p+1} \geq p+1$

it follows then from the definition of σ that $t^\frown\langle k\rangle \preceq \sigma(\beta(n))$; and since this is true for all $n \geq m_{p+1}$ then it follows from the definition of ρ that $t^\frown\langle k\rangle \preceq \rho(\beta(m_{p+1})) \preceq \rho(\beta(n))$. Hence $\beta(m_{p+1})\ R\ \beta(n)$ for all $n \geq m_{p+1}$ and so $\beta(m_{p+1})$ belongs to the distinguished R-chain of β; in particular $|\alpha| > p+1$. On the other hand by the induction hypothesis $\rho(\beta(m_p)) = t$ and since $\beta(m_p)$ belongs to the distinguished R-chain of β then $t \preceq \sigma(\beta(n))$ for all $n \geq m_p$; finally the minimality of m_{p+1} implies $t^\frown\langle k\rangle \not\preceq \sigma(\beta(n))$ for all $n \in]m_p, m_{p+1}[$. It follows from these observations that $\beta(m_{p+1})$ is the successor of $\beta(m_p)$ in α; so $\alpha(p+1) = \beta(m_{p+1})$

and $|\rho(\alpha(p+1))| = p+1$, hence $\rho(\alpha(p+1)) = \rho(\beta(m_{p+1})) = t^\frown\langle k\rangle$, and this finishes the proof of the claim. $\diamond$

In particular it follows from the claim that the distinguished R-chain of any infinite S-branch is infinite, hence S is an expansion of R.

Let $\pi : [R] \to [S]$ be the canonical mapping; we now show that $\pi^{-1}(A_n)$ is $\mathbf{\Delta}^0_1$. For this notice first that the converse of the claim is also true: If $\rho(\alpha(p)) = t$ then $t \in \omega^p$ and $\beta \in F(t)$; this follows from the claim itself and the uniqueness of $t \in \omega^p$ such that $\beta \in F(t)$. Hence for all $t \in \omega^p$ we have: $\pi^{-1}(F(t)) = \bigcup\{N^R_a : h_R(a) = p \text{ and } \rho(a) = t\}$ and so $\pi^{-1}(F(t))$ is open in $[R]$. Moreover since each of the closed sets $A_{(n,k)}$, and in particular $A_n = A_{(n,0)}$, is the union of a subfamily of $(F(s))_{s\in\omega^{n+1}}$ then $\pi^{-1}(A_n)$ (which is already closed by the continuity of π) is also open. $\square$

We recall the following classical notion:

1.6.4. Additively indecomposable ordinals: We recall that an ordinal ξ is said to be <u>additively decomposable</u> if we can find ordinals $\eta < \xi$ and $\zeta < \xi$ such that $\xi = \eta + \zeta$. If not the ordinal is said to be <u>additively indecomposable</u>. Successor ordinals ≥ 2 are obviously additively decomposable. But limit ordinals might be additively decomposable (like $\omega + \omega$) or not (like ω^2). The following elementary result is well known:

LEMMA 1.6.5. *Any additively decomposable limit ordinal ξ admits a decomposition $\xi = \mu + \lambda$ with μ, $\lambda < \xi$ and μ additively indecomposable.*

<u>Proof of Theorem 1.6.2</u>: We prove the statement by induction on $\xi \geq 1$.

For $\xi = 1$ this is just Lemma 1.6.3 above. Suppose the result is true for all $\eta < \xi$, and consider a countable family $(A_n)_{n\in\omega}$ of $\mathbf{\Pi}^0_\xi$ sets in $[S]$. We shall distinguish three cases: 1) ξ is successor; 2) ξ is additively indecomposable; 3) ξ is additively decomposable.

<u>Case 1</u> : *$\xi = \eta + 1$ is successor.*

For all $n \geq 0$ fix a countable family $(A_{(n,k)})_{k\in\omega}$ of $\mathbf{\Pi}^0_\eta$ sets in $[S]$ such that $A_n = \bigcap_{k\in\omega}[S] \setminus A_{(n,k)}$. By the induction hypothesis $\langle S, (A_{(n,k)})_{(n,k)\in\omega^2}\rangle$ is an η-distinguished expansion of some $\langle \tilde{R}, (F_{(n,k)})_{(n,k)\in\omega^2}\rangle$ with $\tilde{R} \subset S$ and each $F_{(n,k)}$ a $\mathbf{\Delta}^0_1$ set in $[\tilde{R}]$. Then for all n, $F_n = \bigcap_{k\geq 0}[R] \setminus F_{(n,k)}$ is a $\mathbf{\Pi}^0_1$ set in $[\tilde{R}]$, and clearly $\langle R, (A_n)_{n\in\omega}\rangle$ is an η-distinguished expansion of $\langle \tilde{R}, (F_n)_{n\in\omega}\rangle$. Then applying Lemma 1.6.3, $\langle \tilde{R}, (F_n)_{n\in\omega}\rangle$ itself is a distinguished expansion of some $\langle R, (G_n)_{n\in\omega}\rangle$ with $R \subset \tilde{R} \subset S$ and each G_n is a $\mathbf{\Delta}^0_1$ set in $[R]$. Then $\langle S, (A_n)_{n\in\omega}\rangle$ is an $\eta + 1 = \xi$-distinguished expansion of $\langle R, (G_n)_{n\in\omega}\rangle$

<u>Case 2</u> : *ξ is additively indecomposable.*

Fix a strictly increasing sequence $(\xi_k)_{k\in\omega}$ of ordinals > 0 converging to ξ, and fix also for all n a sequence $(A_{(n,k)})_{k\in\omega}$ such that for all k, $A_{(n,k)}$ is $\mathbf{\Pi}^0_{\xi_k}$ and $A_n = \bigcap_{k\in\omega} A_{(n,k)}$. Applying the induction hypothesis to each $\xi_k < \xi$, we can ensure that $\langle S, (A_{(n,k)})_{n\in\omega}\rangle$ is an expansion of class ξ_k of some $\langle R_k, (G_{(n,k)})_{n\in\omega}\rangle$ with $R_k \subset S$ and each $G_{(n,k)}$ is $\mathbf{\Delta}^0_1$ in $[R_k]$.

Consider then the infinite product $\tilde{R} = \bigotimes_{k\in\omega} R_k$ (over S): by Theorem 1.5.15 and Theorem 1.5.14 S is a uniform λ-distinguished expansion of $\tilde{R}$, where $\lambda = \sum_{k\in\omega}\xi_k$; but since ξ is additively indecomposable then $\lambda = \xi$. If $\tilde{\pi} : [\tilde{R}] \to [S]$

and $\pi_k : [\tilde{R}] \to [R_k]$ denote the canonical mappings then for all $n \geq 0$, $F_n = \tilde{\pi}^{-1}(A_n) = \bigcap_{k\in\omega} \pi_k^{-1}(G_{(n,k)})$ is $\mathbf{\Pi}^0_1$ in $[\tilde{R}]$, and clearly $\langle S, (A_n)_{n\in\omega}\rangle$ is an ξ-distinguished expansion of $\langle \tilde{R}, (F_n)_{n\in\omega}\rangle$. Then applying Lemma 1.6.3 again, $\langle \tilde{R}, (F_n)_{n\in\omega}\rangle$ is a distinguished expansion of some $\langle R, (G_n)_{n\in\omega}\rangle$ where $R \subset \tilde{R} \subset S$ and each G_n is $\mathbf{\Delta}^0_1$ in $[R]$; hence $\langle S, (A_n)_{n\in\omega}\rangle$ is trivially an $(\xi+1)$-distinguished expansion of $\langle R, (G_n)_{n\in\omega}\rangle$, but Theorem 1.4.3 ensures in fact that $\langle S, (A_n)_{n\in\omega}\rangle$ is a ξ-distinguished expansion of $\langle R, (G_n)_{n\in\omega}\rangle$.

<u>Case 3</u> : *ξ is limit and additively decomposable.*

Applying Lemma 1.6.5 write $\xi = \mu + \lambda$ with $\mu, \lambda < \xi$ and μ additively indecomposable. Notice that since ξ is limit then necessarily λ is also limit. In particular μ and λ are infinite and $1 + \mu = \mu$ and $1 + \lambda = \lambda$.

Since $(A_n)_{n\in\omega}$ is a sequence of $\mathbf{\Pi}^0_{\mu+\lambda}$ sets in $[S]$, there exists a Borel mapping $\varphi : [S] \to \omega^\omega$ of Borel class μ and a sequence $(A'_n)_{n\in\omega}$ of $\mathbf{\Pi}^0_\lambda$ sets, such that $A_n = \varphi^{-1}(A'_n)$ for all n. Fix now any countable basis $(W_k)_{k\in\omega}$ of $\mathbf{\Delta}^0_1$ sets in ω^ω; then for all k, $V_k := \varphi^{-1}(W_k)$ is a $\mathbf{\Delta}^0_\mu$ set in $[S]$, and by the induction hypothesis applied to $\mu < \xi$, $\langle S, (V_k)_{k\in\omega}\rangle$ is a μ-distinguished expansion of some $\langle \tilde{R}, (H_k)_{k\in\omega}\rangle$ with $\tilde{R} \subset S$ and each H_k is $\mathbf{\Delta}^0_1$ in $[\tilde{R}]$. It follows that if $\tilde{\pi} : [\tilde{R}] \to [S]$ denotes the canonical mapping then $\varphi\circ\tilde{\pi} : [\tilde{R}] \to \omega^\omega$ is continuous, hence for all n the set $\tilde{A}_n := \tilde{\pi}^{-1}(A'_n) = \tilde{\pi}^{-1}(\varphi^{-1}(A'_n))$ is $\mathbf{\Pi}^0_\lambda$ in $[\tilde{R}]$. Then by a second application of the induction hypothesis to $\lambda < \xi$, $\langle \tilde{R}, (\tilde{A}_n)_{n\in\omega}\rangle$ is a λ-distinguished expansion of some $\langle R, (G_n)_{n\in\omega}\rangle$ with $R \subset \tilde{R}$ and each G_n is a $\mathbf{\Delta}^0_1$ set in $[R]$. Since $\xi = \mu + \lambda$, it follows that $\langle S, (A_n)_{n\in\omega}\rangle$ is a ξ -distinguished expansion of $\langle R, (G_n)_{n\in\omega}\rangle$ where each G_n is $\mathbf{\Delta}^0_1$ in $[R]$.

This finishes the proof of Theorem 1.6.2. □

The following result which is fundamental for all applications is essentially a by-product of the previous proof, and strictly speaking its statement as well as its proof should have been incorporated to Theorem 1.6.2. However for more clarity we state this result separately and shall only give some indications for the additional arguments needed for the proof.

THEOREM 1.6.6. *a) If $\xi = \eta + 1$ is successor, then any countable family of $\mathbf{\Pi}^0_\xi$ sets is an η-distinguished expansion of $\mathbf{\Pi}^0_1$ sets.*

b) If ξ is limit, then any countable family of $\mathbf{\Pi}^0_\xi$ sets is a uniform ξ-distinguished expansion of $\mathbf{\Pi}^0_1$ sets.

PROOF. A formal proof proceeds by induction. Notice that the initial case $\xi = 1$ is now obvious. Suppose the result is true for all $\eta < \xi$, and consider a countable family $(A_n)_{n\in\omega}$ of $\mathbf{\Pi}^0_\xi$ sets in $[S]$. We also have to distinguish as above three cases: 1) ξ is successor; 2) ξ is additively decomposable; 3) ξ is additively indecomposable.

By a quick inspection one can check that the previous proof already gives in Case 1 that $\langle R, (A_n)_{n\in\omega}\rangle$ is an η-distinguished expansion of some $\langle \tilde{R}, (F_n)_{n\in\omega}\rangle$ with F_n $\mathbf{\Pi}^0_1$ in $[\tilde{R}]$.

Similarly this same proof gives in Case 2 that $\langle R, (A_n)_{n\in\omega}\rangle$ is a uniform ξ-distinguished expansion of some $\langle \tilde{R}, (F_n)_{n\in\omega}\rangle$ with F_n $\mathbf{\Pi}^0_1$ in $[\tilde{R}]$.

For Case 3) the new statement does not appear inside the previous proof. However one can repeat exactly the same arguments up to the construction of

$\tilde{A}_n := \tilde{\pi}^{-1}(A'_n) = \tilde{\pi}^{-1}(\varphi^{-1}(A'_n))$, except for the final step where one now applies the new induction hypothesis to show that $\langle \tilde{R}, (\tilde{A}_n)_{n\in\omega}\rangle$ is a *uniform* λ-distinguished expansion of some $\langle R, (F_n)_{n\in\omega}\rangle$ with each F_n a $\mathbf{\Pi}^0_1$ set in $[R]$. Then as in the previous proof one concludes that $\langle S, (A_n)_{n\in\omega}\rangle$ is a uniform ξ -distinguished expansion of $\langle R, (F_n)_{n\in\omega}\rangle$ where each F_n is a $\mathbf{\Pi}^0_1$ set in $[R]$. □

In the following result we identify as usual the space ω^ω with the corresponding *Ext* space.

THEOREM 1.6.7. *Let $A \subset \omega^\omega$, and μ, ξ, δ be countable ordinals.*

a) If $\xi = \eta + 1$ is successor then:
A is $D_\mu(\mathbf{\Sigma}^0_{1+\xi+\delta})$ iff A is an η-distinguished expansion of a $D_\mu(\mathbf{\Sigma}^0_{1+\delta})$ set.

b) If ξ is limit then:
A is $D_\mu(\mathbf{\Sigma}^0_{1+\xi+\delta})$ iff A is a uniform ξ -distinguished expansion of a $D_\mu(\mathbf{\Sigma}^0_{1+\delta})$ set.

PROOF. The "if" part follows by straightforward computations from Theorem 1.3.8 and Proposition 1.4.1.

For the "only if" part the proof that we sketch briefly, is similar to the argument for Case 3 in the proof of Theorem 1.6.2: Notice that if A is a $D_\mu(\mathbf{\Sigma}^0_{1+\xi+\delta})$ subset of $[S]$, there exists a Borel mapping $\varphi : [S] \to \omega^\omega$ of Borel class ξ and a $D_\mu(\mathbf{\Sigma}^0_{1+\delta})$ set A' in ω^ω such that $A = \varphi^{-1}(A')$. Fix any countable basis $(W_k)_{k\in\omega}$ of $\mathbf{\Delta}^0_1$ sets in ω^ω, so that for all k, $V_k := \varphi^{-1}(W_k)$ is a $\mathbf{\Delta}^0_{1+\xi}$ set in $[S]$. By Theorem 1.6.6, the family $\langle S, (V_k)_{k\in\omega}\rangle$ is an expansion of a family $\langle R, (G_k)_{k\in\omega}\rangle$ with the desired properties and moreover each G_k is a $\mathbf{\Delta}^0_1$ set in $[R]$. It follows that if $\pi : [R] \to [S]$ is the canonical mapping then $\varphi\circ\pi : [R] \to \omega^\omega$ is continuous, hence $\pi^{-1}(A) = \pi^{-1}(\varphi^{-1}(A'))$ is $D_\mu(\mathbf{\Sigma}^0_{1+\delta})$ in $[R]$. □

1.7. Regular expansions and representations

Theorem 1.6.6 relates completely the representation of Borel sets and in this section we shall only give some more practical reformulations motivated by further applications.

In fact as stated Theorem 1.6.6 forces one to treat the Borel sets differently depending on whether their Borel rank is a successor or limit ordinal. For example by Corollary 1.6.7, $\mathbf{\Pi}^0_{\omega+1}$ sets are represented as ω-distinguished expansions of $\mathbf{\Pi}^0_1$ sets, while $\mathbf{\Pi}^0_\omega$ sets are represented as uniform ω-distinguished expansions. This is quite similar to the difficulties one encounters in trying to describe Borel mappings by sequential limits: while Borel mappings of class $\omega + 1$ are exactly all pointwise limits of sequences of Borel mappings of class ω, the analog is not true at level ω, and by a classical result of Lebesgue, any Borel mapping of class $\omega + 1$ is in fact the pointwise limit of a sequence of Borel mappings of finite Baire classes (see [**12**], p.393).

We shall overcome these real and unavoidable difficulties by a purely formal solution, namely by introducing an adequate language which will enable us to restate the results of the previous section in a simpler way.

1.7.1. Regular resolutions and expansions of class ξ. *For any ordinal ξ we define the ordinal ξ^* by:*

$$\begin{cases} 1+\xi = \xi^* + 1 & \text{if } \xi \text{ is successor} \\ \xi = \xi^* & \text{if } \xi \text{ is limit} \end{cases}$$

so

$$\xi^* = \begin{cases} \eta & \text{if } \xi = \eta + 1 \text{ is successor and infinite} \\ \xi & \text{if } \xi \text{ is limit or finite} \end{cases}$$

A <u>regular resolution family of class ξ</u> is a resolution family $(R^{(\eta)})_{\eta \leq \xi^*}$ of order ξ^* which is moreover *uniform* if ξ is *limit*. For example:
– Regular resolution families of class $n < \omega$ are all resolution families $(R^{(k)})_{k \leq n}$ of order n.
– Regular resolution families of class ω are all *uniform* resolution families $(R^{(k)})_{k \leq \omega}$.
– Regular resolution families of class $\omega + 1$ are all resolution families $(R^{(k)})_{k \leq \omega}$.

All notions involving a resolution family admit a "regular version": For example we shall say that S is a <u>regular expansion of class ξ</u> of R, if S is a ξ^*-distinguished expansion of R by a regular resolution family of class ξ . We shall also say that a family $\mathscr{A}$ of subsets of $[S]$ is a <u>regular expansion of class ξ</u> of Γ sets, if $\mathscr{A}$ is a ξ^*-distinguished expansion of Γ sets by a regular resolution family of class ξ ...

Roughly speaking one can view a regular resolution of class ξ of R in S as a coding of some Borel mapping of Baire class ξ from $[S]$ to $[R]$, with *partial domain.* But in all applications we shall consider situations where S is a regular *expansion* of class ξ of R, and in this case this partial mapping will be total:

PROPOSITION 1.7.2. *If S is a regular expansion of class ξ of R then the canonical mapping $\pi : [R] \to [S]$ is a continuous bijection and its inverse $\pi^{-1} : [S] \to [R]$ is of Baire class ξ.*

PROOF. We distinguish the three cases:
– If $\xi = 0$ then $\xi^* = 0$ so $R = S$ and π is the identity mapping.
– If ξ is successor then by Theorem 1.3.8 the image by π of any $\mathbf{\Sigma}^0_1$ subset of $[R]$ is a relatively $\mathbf{\Sigma}^0_{\xi^*+1} = \mathbf{\Sigma}^0_{1+\xi}$ subset of $\pi([R])$.
– If $\xi > 0$ is limit then $\xi^* = \xi$ and R admits a uniform resolution family of order ξ, hence by Proposition 1.4.1 the image by π of any $\mathbf{\Sigma}^0_1$ subset of $[R]$ is a relatively $\mathbf{\Sigma}^0_\xi = \mathbf{\Sigma}^0_{1+\xi}$ subset of $\pi([R])$. □

THEOREM 1.7.3. *For any countable ordinals η, ξ, δ, a subset A of $[S]$ is a $D_\eta(\mathbf{\Sigma}^0_{1+\xi+\delta})$ set iff A is a regular expansion of class ξ of a $D_\eta(\mathbf{\Sigma}^0_{1+\delta})$ set.*

1.7.4. Regular Borel representations of Borel sets: The "only if" part of the previous result will play a central role in all the sequel, and we shall reformulate now its content in complete detail and even with some redundancy :

To any $D_\eta(\mathbf{\Sigma}^0_{1+\xi+\delta})$ subset A of $[S]$ we can associate:

(1) *a subtree R of S such that the canonical mapping $\pi : [R] \to [S]$ is a continuous bijection with inverse mapping $\pi^{-1} : [S] \to [R]$ of Baire class ξ and such that $\hat{A} := \pi^{-1}(A)$ is a $D_\eta(\mathbf{\Sigma}^0_{1+\delta})$ subset of $[R]$.*

(2) *a resolution family* $\overrightarrow{R} = (R^{(\eta)})_{\eta \leq \xi^*}$ *with* $R^{(0)} = S$ *and* $R^{(\xi^*)} = R$ *and*

$$\begin{cases} \xi^* + 1 = 1 + \xi & \text{if } 1 + \xi \text{ is successor} \\ \xi^* = \xi \text{ and } \overrightarrow{R} \text{ is uniform} & \text{if } 1 + \xi = \xi \text{ is limit} \end{cases}$$

We shall then say that $(\hat{A}, R, \overrightarrow{R})$ is a <u>*regular*</u> $D_\eta(\mathbf{\Sigma}^0_{1+\delta})$-<u>*representation of class*</u> ξ of the set $A \subset [S]$.

Substituting everywhere $\check{D}$ to D we define similarly for a $\check{D}_\eta(\mathbf{\Sigma}^0_{1+\xi+\delta})$ subset of $[S]$, the notion of *regular* $\check{D}_\eta(\mathbf{\Sigma}^0_{1+\delta})$*-representation of class* ξ.

For the first applications which we will give in Chapter 3, we will only use regular $\mathbf{\Pi}^0_1$-representations ($\eta = 1$, $\delta = 0$); but for the proof of the main result of the paper in Chapter 6, we will need regular $D(\mathbf{\Sigma}^0_2)$-representations ($\eta = 2$, $\delta = 1$).

CHAPTER 2

A Double-Tree Representation for Borel Sets

The notion of double-tree representation introduced here is a natural continuation of the tree representation of Borel sets considered in Chapter 1. *However aside some basic definitions the results of this chapter will only be needed for the proof of the Main Theorem in* Chapter 6.

Basically the representation of Borel sets developed in the previous chapter enables one to pull back a problem (P) on a Borel set A, say for simplicity $\mathbf{\Pi}^0_n$ with n finite, to a problem (Q) on a simpler Borel set B, say $\mathbf{\Pi}^0_m$ with $m < n$. The way this type of representations can be used in applications might be very elaborate, but will always obey some general principle which in the simplest situations can be stated as follows:

1) Solve the problem (Q) for B

2) Derive the solution of (P) from the solution of (Q).

In this procedure one is tempted naturally to take m minimal, so that step 1) is the simplest possible; and in this respect $\mathbf{\Pi}^0_1$–representations appear to be the "best" representations. However it happens that for the main application that we will give later on, this vague general idea does not apply, and to prove the main result under the optimal hypothesis one has to work with $\mathbf{\Pi}^0_2$-(and even $\mathbf{\Delta}^0_3$)-representations.

Also to exploit such representations we will need to analyze $\mathbf{\Delta}^0_3$ sets in the language of distinguished subrelations. This will be achieved via the notion of double-tree that we will introduce. Though non-trivial, the existence of such an analysis is not really surprising if one keeps in mind that distinguished subrelations code Borel mappings of class one. But more subtle is the combinatorial characterization of $D(\mathbf{\Sigma}^0_2)$ sets which we will prove, and which has no topological analog. This last result will constitute one of the fundamental ingredients of the proof of Main Theorem in Chapter 6.

2.1. Double-trees

EXAMPLE 2.1.1. Let A be a $\mathbf{\Pi}^0_1$ subset of $[R]$, and consider the relations R^+ and R^- defined by:

$$\begin{cases} a\ R^+\ b \iff a\ R\ b \text{ and } (a = 0 \text{ or } a = b \text{ or } N_a \cap A \neq \emptyset) \\ a\ R^-\ b \iff a\ R\ b \text{ and } (a = 0 \text{ or } a = b \text{ or } N_a \cap A = \emptyset) \end{cases}$$

a) It is clear that R^+ and R^- are trees and that $R = R^+ \cup R^-$. Moreover R^+ and R^- are orthogonal in the sense that apart the trivial cases $a = 0$ or $a = b$, we never have simultaneously $a\ R^+\ b$ and $a\ R^-\ b$. Moreover quite obviously both relations are distinguished in R.

b) Finally consider any infinite R-branch $\alpha = (a_n)_{n\geq 0}$: If $\alpha \in A$ then α is also an infinite R^+-branch; and if $\alpha \notin A$ then α contains an infinite R^--branch α', more precisely with $\alpha' = (a_n)_{n\geq p}$ for some p.

The intuition behind these definitions is that the relation $a\ R^+\ b$ is measuring the chances that some infinite branch α passing through b will be in A; and similarly R^- for A^c. Notice also that unlike the algebraic properties described in a), property b) concerns infinite branches and is obviously related, in fact equivalent, to the assumption that A is $\mathbf{\Pi}^0_1$. Nevertheless we shall show that if one reasonably relaxes properties a) and b) above then similar representations exist for more general, namely $\mathbf{\Delta}^0_3$, sets.

We first start by relaxing condition a):

DEFINITION 2.1.2. A double-tree on E is a pair (R^+, R^-) of trees on E satisfying:

1) R^+ and R^- are distinguished subtrees of some tree on E.

2) R^+ and R^- are orthogonal trees, *i.e.*: If $a\ R^+\ b$ and $a\ R^-\ b$ then $a = 0$ or $a = b$.

If R is the smallest tree containing R^+ and R^- we shall say that R is the tree generated by the double-tree (R^+, R^-), or that (R^+, R^-) is a double-tree decomposition of R.

If moreover $R = R^+ \cup R^-$ then we shall say that (R^+, R^-) is a pure double-tree.

REMARKS 2.1.3. a) The double-tree constructed in Example in 2.1.1 is pure, but this property is too restrictive for our purpose. Notice that for an arbitrary double-tree the transitive relation generated by R^+ and R^- is just:

$$R = R^+ \cup R^- \cup (R^+ \circ R^-) \cup (R^- \circ R^+) \cup (R^+ \circ R^- \circ R^+) \cup (R^- \circ R^+ \circ R^-) \ldots\ldots$$

since R^+ and R^- are already tree relations, and condition 1) states implicitly that R is a tree relation; this is a highly non trivial assumption.

b) If R^+ and R^- are distinguished in some tree S then a fortiori they are also distinguished in any intermediate tree, in particular in the tree generated by $R^+ \cup R^-$.

c) If R^+ and R^- are subtrees of some tree S then necessarily R^+ and R^- have the same minimal element as S. In particular the notation 0 in condition 2) is non-ambiguous.

d) We warn the reader that the notion of double-tree introduced here is strictly more general than the one used in [**6**] where we consider only double-trees satisfying the conclusion of Theorem 2.2.2 below.

DEFINITION 2.1.4. Let (R^+, R^-) be a double-tree decomposition of R. We shall say that (R^+, R^-) is a representation double-tree if the sets:

$$A^+ = \{\alpha \text{ contains an infinite } R^+\text{-branch}\}$$
$$A^- = \{\alpha \text{ contains an infinite } R^-\text{-branch}\}$$

form a partition of $[R]$. We shall then say that (R^+, R^-) is a double-tree representation for (A^+, A^-) or simply for A^+.

REMARK 2.1.5. For an arbitrary double-tree decomposition (R^+, R^-) of R, the sets A^+ and A^- defined above are a priori in general position and being a representation double-tree means exactly that the pair (A^+, A^-) forms a partition of the space.

THEOREM 2.1.6. *A admits a pure double-tree representation if and only if A is $\mathbf{\Delta}^0_2$.*

PROOF. Let $A \subset [R]$ admit a pure double-tree representation (R^+, R^-) so satisfy $R = R^+ \cup R^-$. Let $\alpha \in [R]$, and suppose that $\alpha(m)\ R^+\ \alpha(m+1)$ for some m, then for all $n \geq m$ we also have $\alpha(m)\ R^+\ \alpha(n)$: For otherwise by purity we would have $\alpha(m)\ R^-\ \alpha(n)$ hence by distinction $\alpha(m)\ R^-\ \alpha(m+1)$, which is impossible. It follows from this observation that the R^+-distinguished chain of α is the set $\{\alpha(m) :\ \alpha(m)\ R^+\ \alpha(m+1)\}$. Hence by Corollary 1.3.6:

$$\alpha \in A \iff \forall n, \exists m \geq n, \alpha(m)\ R^+\ \alpha(m+1)$$

which shows that A is $\mathbf{\Pi}^0_2$; and similarly for A^c.

The converse implication follows from next more precise result which will be needed later on. We recall that if $R' \subset R \subset S$ and R' is distinguished in S then a fortiori R' is distinguished in R. □

LEMMA 2.1.7. *Let R be a distinguished subtree of S. If $A \subset [R]$ is $\mathbf{\Delta}^0_2$ then A admits a pure double-tree representation (R^+, R^-) such that R^+ and R^- are distinguished in S.*

PROOF. By Hausdorff representation Theorem for $\mathbf{\Delta}^0_2$ sets, there exists some countable ordinal ξ and a family $(F_\eta)_{\eta<\xi}$ of closed subsets of $[R]$ such that if Ω^+ (Ω^-) denotes the set of all ordinals $\leq \xi$ having the opposite (same) parity as ξ, then for all $\alpha \in [R]$:

$$\alpha \in A \iff \min\{\eta < \xi : \alpha \notin F_\eta\} \in \Omega^+$$

where by convention the minimum is ξ whenever $\alpha \in \bigcap_{\eta<\xi} F_\eta$. Then for $\varepsilon \in \{+, -\}$ set:

$$E^\varepsilon = \{0\} \cup \{a \in E :\ \ \mu(a) := \min\{\eta < \xi :\ N_a \cap F_\eta = \emptyset\} \in \Omega^\varepsilon\ \}$$

and define:

$$a\ R^\varepsilon\ b \iff a\ R\ b \ \text{ and } \ a \in E^\varepsilon$$

It is clear that R^+ and R^- are orthogonal subtrees of R and that $R = R^+ \cup R^-$. Suppose that $a\ S\ b\ S\ c$ and $a\ R^+\ c$; then $\mu(a) \in \Omega^+$ and since R is distinguished in S then $a\ R\ b$, hence $a\ R^+\ b$; and similarly for R^-. This proves that R^+ and R^- are distinguished in S, and a fortiori in R, hence (R^+, R^-) is a pure double-tree decomposition of R, and we now prove that it is represents A.

Let α be an infinite R-branch:

– If $\alpha \in A$ then $\zeta := \min\{\eta < \xi : \alpha \notin F_\eta\} < \xi$; so there exists some n such that $N_{\alpha(n)} \cap F_\zeta = \emptyset$, hence $N_{\alpha(m)} \cap F_\zeta = \emptyset$ for all $m \geq n$. On the other hand it follows from the definition of ζ, for all $\eta < \zeta$ and all $m \geq 0$ we have $N_{\alpha(m)} \cap F_\eta \neq \emptyset$. It follows that for all $m \geq n$, $\mu(\alpha(m)) = \zeta$, and since $\alpha \in A$ then $\mu(\alpha(m)) \in \Omega^+$ so $\alpha(m) \in E^+$. This proves that $(\alpha(m))_{m\geq n}$ is an R^+-chain (in fact it is an R^+-branch if we choose n minimal).

– Conversely if $(\alpha(m_k))_{k\geq 0}$ is some infinite R^+-branch then $\mu(\alpha(m_k))_{k\geq 0}$ is a decreasing sequence in Ω^+, hence eventually constant. So there exists some $\zeta \in \Omega^+$

and $j \geq 0$ such that $\mu(\alpha(m_k)) = \zeta$ for all $k \geq j$, then clearly $\zeta = \min\{\eta < \xi : \alpha \notin F_\eta\}$ and $\alpha \in A$.

One proves similarly that $\alpha \notin A$ if and only if α contains an R^--infinite branch. □

We recall that "predecessor" means "*strict predecessor*", and for any tree relation S on E the *S–predecessor mapping* (see 1.1.3) is the mapping $s : E \to E$ defined by:

$$\begin{cases} s(0) = 0 \\ s(a) = \text{the } S\text{-predecessor of } a \quad \text{if } a \neq 0 \end{cases}$$

THEOREM 2.1.8. *The set A admits a double-tree representation if and only if A is $\mathbf{\Delta}^0_3$.*

PROOF. Suppose that A admits a representation by a double-tree (R^+, R^-), and consider the canonical mappings $\pi^+ : [R^+] \to [R]$ and $\pi^- : [R^-] \to [R]$. Then by definition, A is the range of π^+, and A^c is the range of π^-; so by Theorem 1.3.8, A and A^c are both $\mathbf{\Pi}^0_3$.

Conversely suppose that $A \subset S$ is $\mathbf{\Delta}^0_3$ and consider a countable family $(A_j)_{j\geq 0}$ of $\mathbf{\Pi}^0_2$ sets such that $A = \bigcup_{j\geq 0} A_{2j}$ and $A^c = \bigcup_{j\geq 0} A_{2j+1}$. Applying Theorem 1.6.6 we can find a distinguished expansion $\pi : [R] \to [S]$ such that for all j the set $B_j = \pi^{-1}(A_j)$ is $\mathbf{\Pi}^0_1$ in $[R]$, so that $B := \pi^{-1}(A) = \bigcup_{j\geq 0} B_{2j}$ and $B^c = \bigcup_{j\geq 0} B_{2j+1}$ are both $\mathbf{\Sigma}^0_2$, hence B is $\mathbf{\Delta}^0_2$ in $[R]$. Then by Lemma 2.1.7 there exists a double-tree representation (R^+, R^-) of B in $[R]$ such that $R = R^+ \cup R^-$, with moreover R^+ and R^- distinguished in S.

Let r, r^+, r^-, s, denote the predecessor mappings relatively to the tree relations R, R^+, R^-, S, respectively. Notice that $r(a) \in \{r^+(a), r^-(a)\}$ and $r(a)\ R\ s(a)$; and for $a \neq 0$, $r^+(a) \neq r^-(a)$. We now define two functions s^+ and s^- from E to E by:

$$(s^+(a), s^-(a)) = \begin{cases} (r^+(a), r^-(a)) & \text{if } r(a) = s(a) \\ (s(a), r(a)) & \text{if } r(a) \neq s(a) \text{ and } r(a) = r^-(a) \\ |(r(a), s(a)) & \text{if } r(a) \neq s(a) \text{ and } r(a) = r^+(a) \end{cases}$$

Let S^+ and S^- denote the partial orderings generated by the graphs of s^+ and s^-:

$$a\ S^+\ b \iff a = b \text{ or } \exists (a_i)_{0\leq i\leq n} : a_0 = a,\ a_n = b, \text{ and } \forall i < n,\ a_i = s^+(a_{i+1})$$

$$a\ S^-\ b \iff a = b \text{ or } \exists (a_i)_{0\leq i\leq n} : a_0 = a,\ a_n = b, \text{ and } \forall i < n,\ a_i = s^-(a_{i+1})$$

Notice that for all a we have $s^+(a) \in \{s(a), r^+(a)\}$ hence $s^+(a)\ S\ a$; similarly $s^-(a)\ S\ a$. It follows that S^+ and S^- are subtrees of S with s^+ and s^- as predecessor mappings. Moreover since for all a, $s(a)$ is equal either to $s^+(a)$ or to $s^-(a)$, then S is generated by $S^+ \cup S^-$.

We shall now establish a number of properties of S^+ and S^-; and since the situation is symmetrical we state them only for S^+.

CLAIM 1. $R^+ \subset S^+$

Proof: Assume that $a\ R^+\ b$; and let $(b_k)_{k\leq n}$ be the decreasing enumeration of the S^+–branch ending at b, so $b_0 = b$, $b_{k+1} = s^+(b_k)$, $b_n = 0$. Let $k \leq n$ be the largest integer such that $a\ S\ b_k$. If $a = b_k$ then $(b_k, b_{k-1}, \ldots, b_0)$ witnesses that $a\ S^+\ b$. If not since $a\ R^+\ b_0$ then by the distinction of R^+ in S we have

$a\ R^+\ b_k$, hence $a\ R^+\ r^+(b_k)$ and a fortiori $a\ S\ r^+(b_k)$. And since $a\ S\ b_k$ we also have $a\ S\ s(b_k)$. Finally since $s^+(b_k) \in \{s(b_k), r^+(b_k)\}$, we have $a\ S\ s^+(b_k) = b_{k+1}$ which contradicts the maximality of k. ◇

CLAIM 2. *S^+ is distinguished in S*

Proof: Suppose that $a\ S\ b\ S\ c$ and $a\ S^+\ c$; we have to show that $a\ S^+\ b$; and without loss of generality we can assume that $b = s(c)$. Consider then $b' := r^+(c)$: we have $b'\ R^+\ c$ thus $b'\ R^+\ b$ since R^+ is distinguished in S, hence $b'\ S^+\ b$ by Claim 1. Since $a\ S^+\ c$ and $a \neq c$ we have $a\ S^+\ s^+(c)$ and $s^+(c) \in \{s(c), r^+(c)\} = \{b, b'\}$. If $s^+(c) = b$ then $a\ S^+\ b$. And if $s^+(c) = b'$ then $a\ S^+\ b'\ S^+\ b$, hence $a\ S^+\ b$ by transitivity. ◇

CLAIM 3. *S^+ and S^- are orthogonal.*

Proof: Suppose that for $a \neq b$ we have $a\ S^+\ b$ and $a\ S^-\ b$; then by the distinction of S^+ and S^- in S we also have $a\ S^+\ a'$ and $a\ S^-\ a'$ for all a' such that $a\ S\ a'\ S\ b$; in particular if a' is the S-successor of a under b we obtain: $a = s(a') = s^+(a') = s^-(a')$ hence $r^+(a) = r^-(a)$, and this implies that $a = 0$ since R^+ and R^- are orthogonal. ◇

CLAIM 4. $R^+ = R \cap S^+$

Proof: Write: $R \cap S^+ = (R^+ \cup R^-) \cap S^+ = (R^+ \cap S^+) \cup (R^- \cap S^+)$ and notice that by Claim 2 and Claim 3: $(R^+ \cap S^+) = R^+$ and $R^- \cap S^+ \subset S^- \cap S^+$ is the trivial tree. ◇

CLAIM 5. *No $\alpha \in [S]$ contains an infinite R^+-chain and an infinite R^--chain.*

Proof: Suppose that $\alpha \in [S]$ contains an infinite R^+-chain $(\alpha(m))_{m \in M}$ and an infinite R^--chain $(\alpha(p)_{p \in P}$, we shall prove that $\beta = (\alpha(n))_{n \in M \cup P}$ is an infinite R-chain.

So pick m and p in $M \cup P$; we have to show that $\alpha(m)\ R\ \alpha(p)$ or $\alpha(p)\ R\ \alpha(m)$. If both indices are either in M or in P this is obvious. So suppose that $m \in M$ and $p \in P$. If $m < p$, then since M is infinite we can pick $m' \in M$ such that $\alpha(m)\ S\ \alpha(p)\ S\ \alpha(m')$, and since $\alpha(m)\ R^+\ \alpha(m')$ then by the distinction of R^+ in S we also have $\alpha(m)\ R^+\ \alpha(p)$, hence $\alpha(m)\ R\ \alpha(p)$; and similarly if $p < m$.

Recall that (R^+, R^-) is a representation for $B = \pi^{-1}(A)$. So if $\gamma \in [R]$ is the unique infinite R-branch containing β. then γ contains an infinite R^+-chain and an infinite R^--chain, hence $\gamma \in B \cap B^c$ which gives the contradiction. ◇

We now come to the main property:

CLAIM 6. *No $\alpha \in [S]$ contains an infinite S^+-chain and an infinite R^--chain.*

Proof: Again suppose by contradiction that α contains an infinite S^+-branch $(\alpha(m))_{m \in M}$, and an infinite R^--chain $(\alpha(p)_{p \in P}$; we shall then prove that α contains an infinite R^+-chain, which by Claim 5 will give the contradiction.

Notice that if $m > n > 0$ were both in $M \cap P$ then we would have by Claim 1: $\alpha(n)\ S^+\ \alpha(m)$ and $\alpha(n)\ S^-\ \alpha(m)$ which is impossible; hence $M \cap P$ contains at most one non zero element.

Let $(m_k)_{k \geq 0}$ be the canonical enumeration of M defined by $h_{S^+}(\alpha(m_k)) = k$, and consider the set

$$M' = \{m_k \in M : \exists p > m_k ,\ p \in P \text{ and } \alpha(m_k)\ S\ \alpha(p)\ S\ \alpha(m_{k+1})\}$$

Since both M and P are infinite and almost disjoint then clearly M' is also infinite; and we shall now prove that $(\alpha(m))_{m \in M'}$ is an infinite R^+-chain.

Let $m_j < m_k$ in M', set $q = m_{j+1}$ and fix $p > m_j$ in P such that

$$\alpha(m_j)\ S\ \alpha(p)\ S\ \alpha(q)\ S\ \alpha(m_k)$$

We have to show that $\alpha(m_j)\ R^+\ \alpha(m_k)$.

First pick any $p' > m_k$ in P; then $\alpha(p)\ S\ \alpha(m_k)\ S\ \alpha(p')$ and since $\alpha(p)\ R^-\ \alpha(p')$ then by the distinction of R^- in S we have $\alpha(p)\ R^-\ \alpha(m_k)$.

Second, recall that $h_{S^+}(\alpha(m_j)) = m_j$ and $h_{S^+}(\alpha(q)) = m_j + 1$ hence $\alpha(m_j)$ is the S^+-predecessor of $\alpha(q)$ but is not its S-predecessor (since $\alpha(p)$ is in-between). It follows then from the definition of the s^+ mapping that $r^+(\alpha(q)) = s^+(\alpha(q)) = \alpha(m_j)$, hence $\alpha(m_j)\ R^+\ \alpha(q)$; and again by he distinction of R^+ in S we have $\alpha(m_j)\ R^+\ \alpha(p)$ also.

It follows from these observations that $\alpha(m_j)\ R^+\ \alpha(p)\ R^-\ \alpha(m_k)$ hence $\alpha(m_j)\ R\ \alpha(m_k)$. Finally since by assumption we also have $\alpha(m_j)\ S^+\ \alpha(m_k)$ then from Claim 4 we can infer that $\alpha(m_j)\ R^+\ \alpha(m_k)$. ◇

To finish notice that from the previous claims (S^+, S^-) is clearly a double-tree generating the tree S; moreover for any $\alpha \in [S]$:

– If $\alpha \in A$ then $\beta = \pi^{-1}(\alpha) \in \pi^{-1}(A) \subset [R]$ is an infinite R-branch contained in α. Since (R^+, R^-) is a representation for $\pi^{-1}(A)$ then β contains some infinite R^+-branch γ which by Claim 1 is an infinite S^+-chain. Hence α contains an infinite S^+-branch.

– Conversely if α contains an infinite S^+-branch then $\alpha \in A$, for otherwise by the same argument as above (applied to A^c) α would contain some infinite R^--chain contradicting Claim 6.

Hence (S^+, S^-) is a representation for A and this ends the proof of Theorem 2.1.8. □

2.2. Double-tree characterization of Σ_2^0 sets

The notion of double-tree representation is symmetrical and does not a priori distinguish the represented set from its complement. However as we shall see next this is no more the case if one takes into account the relative positions of the R^+-chains and R^--chains.

2.2.1. Intertwining of the double chains in a double-tree: Let (R^+, R^-) be a double-tree (non necessarily a representation) on a set E. For any element $a \in E$ one can consider :

$\sigma_{R^+}(a) = (a_0^+, a_1^+, \ldots, a_p^+)$: the R^+-chain of all R^+-predecessors of a.

$\sigma_{R^-}(a) = (a_0^-, a_1^-, \ldots, a_q^-)$: the R^--chain of all R^--predecessors of a.

Notice that on $\sigma_{R^+}(a)$ the order induced by R^+ is the same as the order induced by R, and similarly for $\sigma_{R^-}(a)$. Also both $\sigma_{R^+}(a)$ and $\sigma_{R^-}(a)$ are subsequences of

$\sigma_R(a) = (a_0, a_1, \ldots, a_n)$: the R-chain of all R-predecessors of a.

In particular the elements a_i^+ are R-comparable with the elements a_j^-; and the two sequences $\sigma_{R^+}(a)$ and $\sigma_{R^-}(a)$ can a priori intertwine (relatively to R) in an arbitrary way. Notice however that $a_0^+ = a_0^- = 0$, and by the orthogonality condition 0 is the unique common element to $\sigma_{R^+}(a)$ and $\sigma_{R^-}(a)$. Also it will be more convenient to introduce the <u>strict</u> chains:

$\sigma^*_{R^+}(a) = (a_1^+, \dots, a_p^+)$: the strict R^+-chain of all non-zero R^+-predecessors of a.

$\sigma^*_{R^-}(a) = (a_1^-, \dots, a_q^-)$: the strict R^--chain of all non-zero R^--predecessors of a.

For example in the double-tree of Example 2.1.1 we always have:

$$a_1^+ \; R \; a_2^+ \; R \dots a_{p-1}^+ \; R \; a_p^+ \; R \; a_1^- \; R \; a_2^- \; R \dots\dots a_{q-1}^- \; R \; a_q^-$$

Thus in this double-tree the strict R^+-chain of a given element occurs before its strict R^--chain. Notice by the way that in this particular case since $R = R^+ \cup R^-$ then we also have:

$$\sigma^*_R(a) = \sigma^*_{R^+}(a) \cup \sigma^*_{R^-}(a)$$

but in general $\sigma^*_R(a)$ might contain elements which appear neither in $\sigma^*_{R^+}(a)$ nor in $\sigma^*_{R^-}(a)$.

THEOREM 2.2.2. *A subset of $[R]$ is $\mathbf{\Sigma}_2^0$ iff it admits a double-tree representation (R^+, R^-) such that the strict R^+-chain of any element a is entirely between a and its R^--predecessor.*

This will follow from next two more precise lemmas which we will need later on.

LEMMA 2.2.3. *Let A be a $\mathbf{\Sigma}_2^0$ subset of $[R]$. Then it admits a double-tree representation (R^+, R^-) such that the strict R^+-chain of any element a is entirely between a and its R^--predecessor.*

Moreover if h denotes the R^--height function then:

1) $a \, R \, b \implies h(a) \leq h(b)$

2) $a \, R^+ \, b \iff \big(a = 0 \;\text{ or }\; (a \, R \, b \;\text{ and }\; h(a) = h(b)\,)\big)$

PROOF. Suppose that A is $\mathbf{\Sigma}_2^0$ in $[R]$ and fix a sequence $(F_n)_{n \geq 0}$ of closed sets in $[R]$ such that $F_0 = \emptyset$ and $A = \bigcup_{n \geq 0} F_n$. For any $a \in \text{dom}(R)$ with $a \neq 0$ let a^* denote the R-predecessor of a, and define by induction on R

$$h(a) = \begin{cases} 0 & \text{if } a = 0 \\ \max\{n < h(a^*) + 1 : \; N_a \cap F_k = \emptyset \text{ for all } k < n\} & \text{if } a \neq 0 \end{cases}$$

Since $F_0 = \emptyset$ then $h(a) = 0$ iff $a = 0$; and for $a \neq 0$ it is clear that $h(a^*) \leq h(a) \leq h(a^*) + 1$; in particular if $a \; R \; b$ then $h(a) \leq h(b)$. Moreover for all $\alpha \in [R]$ the sequence $(h(\alpha(n)))_{n \geq 0}$ satisfies $h(\alpha(n)) \leq h(\alpha(n+1)) \leq h(\alpha(n)) + 1$ for all n, and clearly:

$$\alpha \in A \iff \sup\{h(\alpha(n)) \; : \; n \geq 0\} < \infty$$

We define the relations R^+ and R^- by:

$$a \; R^+ \; b \iff a = 0 \text{ or } (a \; R \; b \text{ and } h(a) = h(b))$$

$$a \; R^- \; b \iff a = b \text{ or } (a \; R \; b \text{ and } h(a) < h(\sigma(a,b)))$$

where $\sigma(a, b)$ (defined only if $a \; R \; b$ and $a \neq 0$) is the R-successor of a below b.

It is straightforward to check that R^+ is a distinguished tree relation in R. It is also clear that R^- is a partial order relation, and is distinguished in R (notice that if $a\ R\ b\ R\ c$ with a distinct from b, then $\sigma(a,b) = \sigma(a,c)$); hence R^- is a tree relation in R. Moreover R^+ and R^- are obviously orthogonal and so (R^+, R^-) is a double-tree.

For any $a \in \mathrm{dom}(R)$ with $a \neq 0$, either $h(a^*) = h(a)$ and then $a^*\ R^+\ a$; or else $h(a^*) < h(a)$, and since $a = \sigma(a^*, a)$ then in this case $a^*\ R^-\ a$. So $a^*\ R^+ \cup R^-\ a$; and this proves that R is the tree generated by R^+ and R^-.

Let $\alpha \in [R]$:

– If $\alpha \in A$ then there exist m and k such that $h(\alpha(n)) = k$ for all $n \geq m$ hence by definition of R^+ $(\alpha(n))_{n \geq m}$ is an infinite R^+-chain contained in α.

– If $\alpha \notin A$ then the set $M = \{m :\ h(\alpha(m)) < h(\alpha(m+1))\}$ is infinite. Then for all $m < m'$ in M we have $\sigma(\alpha(m), \alpha(m')) = \alpha(m+1)$ hence by definition of R^-, $\alpha(m)\ R^-\ \alpha(m')$. So $\big(\alpha(m)\big)_{m \in M}$ is an infinite R^--chain contained in α.

Moreover it is clear from the definitions of R^+ and R^- that α never contains simultaneously an infinite R^+-chain and an infinite R^--chain. This proves that (R^+, R^-) is a double-tree representation for the set A.

For any $a \in \mathrm{dom}(R)$, if a' is any R^--predecessor of a, and a'' is any non-zero R^+-predecessor of a, then $h(a') < h(a) = h(a'')$ hence necessarily $a'\ R\ a''$. This proves that the strict R^+-chain of a is entirely between a and its R^--predecessor.

To finish notice that the function h satisfies obviously properties 1) and 2), and one easily checks that h is precisely the height function of the tree relation R^-. □

LEMMA 2.2.4. *If $A \subset [R]$ admits a double-tree representation (R^+, R^-) such that the strict R^+-chain of any element a is entirely between a and its R^--predecessor, then:*

$$A = \{\alpha \in [R] :\ \exists n, \forall m \geq n,\ \alpha(m)\ R^+\ \alpha(m+1)\}$$

in particular A is $\boldsymbol{\Sigma}^0_2$.

PROOF. Let $\alpha \in A$ and let $(\alpha(p))_{p \in P}$ be the (unique) infinite R^+-branch contained in α.

CLAIM . *The set $M = \{m :\ \alpha(m)\ R^-\ \alpha(m+1)\}$ is finite.*

Proof: Otherwise we can find integers $p < m < m+1 < q$ such that $m \in M$, p and q in P; so we have $\alpha(p)\ R\ \alpha(m)\ R^-\ \alpha(m+1)\ R\ \alpha(q)$ with $\alpha(p)\ R^+\ \alpha(q)$, and by the distinction of R^+ we also have $\alpha(p)\ R^+\ \alpha(m+1)$. It follows that $\alpha(p)$ is an R^+-predecessor of $a := \alpha(m+1)$ and $\alpha(m)$ is the R^--predecessor of a, and since $p < m$ this contradicts the hypothesis, and proves the Claim. ◇

Since R is generated by R^+ and R^- then for all $m > 0$, either $\alpha(m)\ R^+\ \alpha(m+1)$ or $\alpha(m)\ R^-\ \alpha(m+1)$; hence by the Claim there exists some n such that for all $m \geq n$, $\alpha(m)\ R^+\ \alpha(m+1)$. This shows one inclusion; the converse inclusion is trivial. □

REMARK 2.2.5. By completely similar arguments one can prove the following variation of Theorem 2.2.2:

A subset of $[R]$ is $\boldsymbol{\Sigma}^0_1$ iff it admits a pure *double-tree representation (R^+, R^-) such that the strict R^+-chain of any element a is entirely between a and its R^--predecessor.*

2.3. Double-tree characterization of $\mathbf{D}(\boldsymbol{\Sigma}^0_2)$ sets

THEOREM 2.3.1. *A subset of $[R]$ is $D(\boldsymbol{\Sigma}^0_2)$ iff it admits a double-tree representation (R^+, R^-) such that the strict R^+-chain of any element a is entirely between two consecutive elements of the R^--chain of a.*

PROOF. The proof is quite long and we split it into two parts:

Part I: *The condition is necessary.*

Suppose that A is $D(\boldsymbol{\Sigma}^0_2)$ in $[R]$, and write $A = B \cap C$ with B $\boldsymbol{\Pi}^0_2$ and C $\boldsymbol{\Sigma}^0_2$. Fix two double-tree representations $\mathscr{S} = (S^+, S^-)$ and $\mathscr{T} = (T^+, T^-)$ for the $\boldsymbol{\Sigma}^0_2$ sets $[R] \setminus B$ and C, given by Lemma 2.2.3. Let h denote the height function of S^-, and $|\ .\ |$ denote the height function of R; we recall that:

(1) $a\, R\, b \implies h(a) \leq h(b)$

(2) $a\, S^+\, b \iff a = 0$ or $(a\, S\, b$ and $h(a) = h(b)\,)$

Notice that for all a we also have $h(a) \leq |a|$, and consider the mapping $\varphi : E \to E$ uniquely defined by:

$$\varphi(a)\ R\ a \qquad \text{and} \qquad |\varphi(a)| = h(a)$$

CLAIM 1. *If $a\, R\, b$ then $\varphi(a)\ R\ \varphi(b)$. Moreover φ realizes an increasing bijection from the S^--branch of a onto the R-branch of $\varphi(a)$.*

Proof: If $a\, R\, b$ then $\varphi(a)$ and $\varphi(b)$ are R-predecessors of b, hence R-comparable. Moreover by (1), $|\varphi(a)| = h(a) \leq h(b) = |\varphi(b)|$ and so $\varphi(a)\ R\ \varphi(b)$.

The second part of the conclusion follows from the relation $|\varphi(a)| = h(a)$. ◇

Now we define the relations R^+ and R^- by:

$$\begin{cases} a\, R^+\, b \iff a\, S^-\, b \text{ and } \varphi(a)\ T^+\ \varphi(b) \\ \\ a\, R^-\, b \iff a\, R\, b \text{ and } \varphi(a)\ T^-\ \varphi(b) \end{cases}$$

CLAIM 2. *$\mathscr{R} = (R^+, R^-)$ is a double-tree.*

Proof: It is clear that R^+ and R^- are partial orders and finer than R. If $a\, R\, b\, R\, c$ then by Claim 1, $\varphi(a)\ R\ \varphi(b)\ R\ \varphi(c)$; it follows then directly from the distinction of S^- and T^+ in R that if $a\, R^+\, c$ then $a\, R^+\, b$. This proves that R^+ is distinguished in R, hence by Proposition 1.2.3 R^+ is a distinguished subtree of R; and similarly for R^-.

To finish we now prove that R^+ and R^- are orthogonal: So suppose that $a\, R^+\, b$ and $a\, R^-\, b$ for some elements a and b, then $\varphi(a)\ T^+\ \varphi(b)$ and $\varphi(a)\ T^-\ \varphi(b)$ hence by the orthogonality of (T^+, T^-) we have $\varphi(a) = \varphi(b)$ so, from the definition of φ, $h(a) = h(b)$; but since we also have $a\, R\, b$ then by property (2), we have $a\, S^+\, b$. On the other hand from $a\, R^+\, b$, we can infer $a\, S^-\, b$; hence by the orthogonality of (S^+, S^-) we can conclude that $a = 0$ or $a = b$. ◇

CLAIM 3. *The tree generated by $\mathscr{R}$ is R.*

Proof: We have to prove that for any $a \neq 0$ if a^* denotes the R-predecessor of a then either $a^*\, R^+\, a$ or $a^*\, R^-\, a$. So we fix such an element a, and notice that by Claim 1, $\varphi(a^*)\ R\ \varphi(a)$. Since (S^+, S^-) generates R we can distinguish two cases:

– If $a^*\, S^+\, a$ then by (2) $h(a^*) = h(a)$, so $|\varphi(a^*)| = |\varphi(a)|$ hence $\varphi(a^*) = \varphi(a)$, and consequently $a^*\, R^-\, a$.

– If $a^* \ S^- \ a$ then again by property (2) $h(a^*) = h(a) - 1$, so $|\varphi(a^*)| = |\varphi(a)| - 1$ hence $\varphi(a^*)$ is the R-predecessor of $\varphi(a)$; but since (T^+, T^-) also generates R then: either $\varphi(a^*) \ T^+ \ \varphi(a)$ and in this case we have $a^* \ R^+ \ a$ (since $a^* \ S^- \ a$), or $\varphi(a^*) \ T^- \ \varphi(a)$ and in this case we have $a^* \ R^- \ a$. ◇

CLAIM 4. *$\mathscr{R}$ is a representation double-tree for A.*

Proof: Consider some $\alpha \in [R]$:

– Suppose that $\alpha \in A = B \cap C$; since $\alpha \in B$ it contains some infinite S^--chain $(\alpha(m_j))_{j \geq 0}$; then by Claim 1, $(\varphi(\alpha(m_j))_{j \geq 0}$ is also an R-chain, and since $\alpha \in C$ then $(\varphi(\alpha(m_j))_{j \geq 0}$ contains some infinite T^+-chain $(\varphi(\alpha(m_{j_k})))_{k \geq 0}$. It follows then from the definitions that $(\alpha(m_{j_k}))_{k \geq 0}$ is an R^+-chain.

– Suppose that $\alpha \in B \setminus C$; then by exactly the same arguments as in the previous case one shows that α contains some infinite T^--chain $(\varphi(\alpha(m_{j_k})))_{k \geq 0}$, hence $(\alpha(m_{j_k}))_{k \geq 0}$ is an R^--chain.

– Suppose that $\alpha \notin B$; then it contains some infinite S^+-chain $(\alpha(m_j))_{j \geq 0}$; then by property (2), $\varphi(\alpha(m_j)) = a$ is constant. It follows then from the definitions that $(\alpha(m_j))_{j \geq 0}$ is an R^--chain.

Finally we now show that no $\alpha \in [R]$ contains simultaneously an infinite R^+-chain and an infinite R^--chain: for otherwise α would contain in particular an infinite S^--chain, and taking its image by φ we would get an infinite R-chain say β; but it follows then from the assumptions on α that β also would contain simultaneously an infinite T^+-chain and an infinite T^--chain which is impossible, since (T^+, T^-) is a representation double-tree. This proves that $\mathscr{R}$ is a representation double-tree for A. ◇

CLAIM 5. *The strict R^+-chain of any element a is entirely between two consecutive elements of the R^--chain of a.*

Proof: If $h_{R^+}(a) \leq 1$ the strict chain is empty. So suppose that $h_{R^+}(a) \geq 2$ and suppose that the unique element a' defined by $a' \ R^+ \ a$ and $h_{R^+}(a') = 1$ satisfies $b \ R \ a' \ R \ c$ for some $b \ R^- \ c \ R^- \ a$; we have to show that any element a'' of the strict R^+-chain of a satisfies also $b \ R \ a'' \ R \ c$.

Since $b \ R \ a' \ R^+ \ a''$ then obviously $b \ R \ a''$. Moreover since $a' \ R^+ \ a$ then $\varphi(a') \ T^+ \ \varphi(a)$; and by (1) we have $\varphi(a') \ R \ \varphi(c) \ R \ \varphi(a)$, but since $\mathscr{T}$ satisfies Proposition 2.2.2, then we necessarily have $\varphi(a') \ T^+ \ \varphi(c) \ T^+ \ \varphi(a)$. On the other hand since $c \ R^- \ a$ we also have $\varphi(c) \ T^- \ \varphi(a)$; hence $\varphi(c) = \varphi(a)$. Similarly we have $\varphi(a'') \ R \ \varphi(c) = \varphi(a)$, and if we had $c \ R \ a''$ then we would also have $\varphi(a'') = \varphi(c) = \varphi(a)$, which is impossible by Claim 1, since a'' is a strict S^--predecessor of a. Hence $b \ R \ a'' \ R \ c$. ◇

This finishes the proof of Part I.

Part II: *The condition is sufficient.*

Let (R^+, R^-) be an arbitrary double-tree generating R and consider the sets

$$A^+ = \{\alpha \in [R] : \text{ contains an infinite } R^+\text{-branch}\}$$
$$A^- = \{\alpha \in [R] : \text{ contains an infinite } R^-\text{-branch}\}$$

which we recall are not necessarily disjoint or covering $[R]$. Define the sets G, H by:

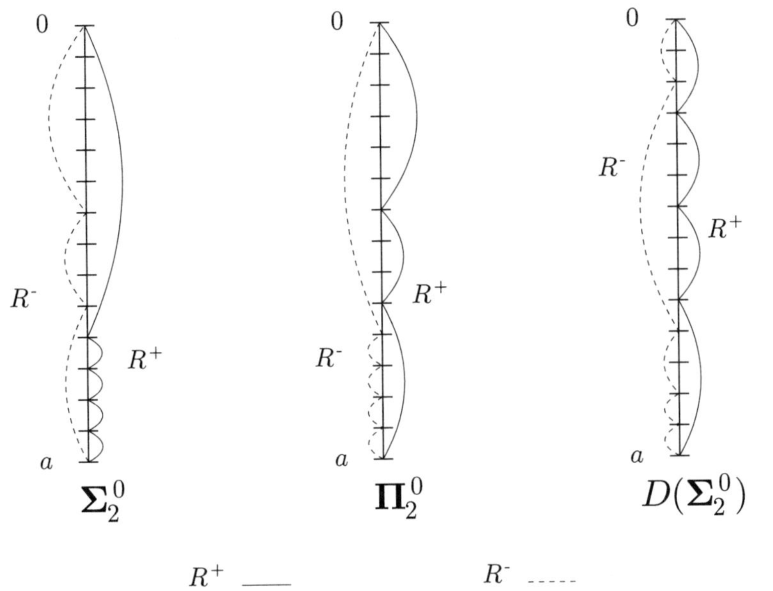

FIGURE 2.1. The pictures above describe the intertwining of the R^+-chain and R^--chain along the R-chain of an element a when the double-tree (R^+, R^-) or its dual is given by Theorem 2.2.2 (first two pictures), or by Theorem 2.3.1 (last picture). Notice that the second example is obtained from the first one by swaping the pair (R^+, R^-), and both are degenerate cases of the third example.

$$G = \{\alpha \in [R] : \ \forall n, \exists m \geq n, \ \alpha(m) \ R^+ \ \alpha(m+1)\}$$

$$H = \{\alpha \in [R] : \ \exists n, \forall q > p \geq n, \quad \alpha(p) \ R^- \ \alpha(q) \to \forall m \in [p, q[, \ \alpha(m) \ R^- \ \alpha(m+1)\}$$

By straightforward computations G is $\mathbf{\Pi}_2^0$, and H is $\mathbf{\Sigma}_2^0$.

CLAIM 6. $G \cap H \subset (A^-)^c$

Proof: We shall show that $(A^- \cap G) \subset H^c$: If $\alpha \in A^- \cap G$ let $(\alpha(m_j))_{j \geq 0}$ be an infinite R^--chain contained in α. Fix $n = m_{j_0}$ for an arbitrary j_0; since $a \in G$ we can find $m \geq n$ such that $\alpha(m) \ R^+ \ \alpha(m+1)$. Let j defined by $m_j \leq m < m_{j+1}$, then $p := m_j$ and $q := m_{j+1}$ are such that $q > p \geq n$ and $\alpha(p) \ R^- \ \alpha(q)$ but $\alpha(m) \ R^+ \ \alpha(m+1)$ for some $m \in [p, q[$. This proves that $\alpha \notin H$. $\diamond$

Claim 7. *If the strict R^+-chain of any element a is entirely between two consecutive elements of the R^--chain of a, then $A^+ \subset G \cap H$*

Proof: Obviously $A^+ \subset G$, and we shall now prove that $A^+ \subset H$.

So consider $\alpha \in A^+$ and let $(\alpha(m_j))_{j\geq 0}$ be an infinite R^--chain contained in α. We argue by contradiction: If $\alpha \notin H$ then for $n = m_1 > 0$ we can find $q > p \geq m_1$ and $m \in [p, q[$ such that $\alpha(p)\ R^-\ \alpha(q)$ and $\alpha(m)\ R^+\ \alpha(m+1)$. It follows then from the distinction of R^- that $\alpha(p)\ R^-\ \alpha(m+1)$, hence we can suppose that $q = m+1$ so that $\alpha(m)$ is now the R-predecessor of $\alpha(q)$. On the other hand since $\alpha(m_1)$ is in the distinguished chain of α then necessarily $\alpha(m_1)\ R^+\ \alpha(m)$. Hence we have $\alpha(m_1)\ R\ \alpha(p)\ R\ \alpha(m)\ R\ \alpha(q)$ with $\alpha(m_1)\ R^+\ \alpha(m)\ R^+\ \alpha(q)$ and $\alpha(p)\ R^-\ \alpha(q)$ and this contradicts the intertwining condition on the double chain of $\alpha(q)$. ◇

To finish the proof of Part II, notice that if (R^+, R^-) is a representation double-tree then $(A^-)^c = A^+$ hence under the intertwining condition $A^+ = G \cap H$ and so is necessarily a $D(\mathbf{\Sigma}^0_2)$ set. □

2.4. Appendix : extension to Wadge classes

We shall discuss now briefly an alternative description of Wadge classes inspired by the double-tree representation. The results of this section are unessential for the rest of the paper and will only be used in some marginal parts.

2.4.1. Wadge classes. Recall that a pointclass Γ is said to be a Wadge class if it has a generator, that is if there exists some set $A_0 \subset 2^\omega$ such that $\Gamma = \{\varphi^{-1}(A_0) : \varphi \text{ continuous}\}$. If Γ is a Wadge class then clearly its dual class $\check{\Gamma}$ is also a Wadge class, and we set as usual $\Delta(\Gamma) := \check{\Gamma} \cap \Gamma$.

If $\check{\Gamma} = \Gamma$ then Γ is said to be *self-dual*. A well-known result of Martin, gives an explicit description of self-dual Wadge classes from non-self-dual classes; and for simplicity we shall restrict our discussion here to the latter case.

The simplest Wadge classes are $\{\emptyset\}$ and its dual class $\{\check{\emptyset}\}$. Less trivial Wadge classes are provided by the Baire classes $\mathbf{\Sigma}^0_\xi$, $\mathbf{\Pi}^0_\xi$ and more generally the difference classes $D_\eta(\mathbf{\Sigma}^0_\xi)$. A complete description of Wadge classes of Borel sets was given by Wadge in [**27**] a long time ago. This description makes use of two types of operations: 1) preimages by Borel mappings; 2) a number of specific Boolean operations (see [**13**] for more details). In [**15**] the authors give a formally simpler description using a unique scheme of operations $\mathbf{S}_\xi(\Gamma, \Gamma')$ with $\xi \in \omega_1$. In fact we shall consider a slightly modified version of this description, which is more adapted to our purpose, and that we present now.

2.4.2. The operations EXP and GSU. :

- For any set A and any countable ordinal ξ, consider the class:
$$\mathbf{EXP}^{(\xi)}(A) := \{\varphi^{-1}(A) : \quad \varphi \text{ a Borel mapping of class } \xi\}$$
- For any sequence $(A_n)_{n\geq 0}$ of sets we consider the class:

$$\mathbf{GSU}\big((A_n)_{n\geq 0}\big) : \left\{ \begin{array}{l} \left(\bigcup_{n>0} A_n \cap U_n\right) \cup \left(A_0 \setminus \bigcup_{n>0} U_n\right) \text{ where } (U_n)_{n>0} \\ \text{is any sequence of pairwise disjoint basic clopen sets} \end{array} \right\}$$

More generally for any class Γ and any sequence $(\Gamma_n)_{n\geq 0}$ of classes we set:

$$\mathbf{EXP}^{(\xi)}(\Gamma) := \bigcup\{\mathbf{EXP}^{(\xi)}(A) : \quad A \in \Gamma\}$$
$$\mathbf{GSU}\big((\Gamma_n)_{n\geq 0}\big) := \bigcup\{\mathbf{GSU}\big((A_n)_{n\geq 0}\big) : \quad A_n \in \Gamma_n \text{ for all } n\}$$

We shall say that $\mathbf{EXP}^{(\xi)}(\Gamma)$ is the <u>ξ-expansion</u> of Γ and $\mathbf{GSU}\big((\Gamma_n)_{n\geq 0}\big)$ is the <u>generalized separated union</u> of the family of classes $\big((\Gamma_n)_{n\geq 0}\big)$.

These operations are closely related to the operations $\mathbf{S}_\xi$ of [**15**]; more precisely one can easily check that $\mathbf{S}_{1+\xi}(\Gamma, \Gamma') = \mathbf{EXP}^{(\xi)}(\mathbf{S}_1(\Gamma, \Gamma'))$ and $\mathbf{GSU}\big((\Gamma_n)_{n\geq 0}\big) = \mathbf{S}_1(\bigcup_{n>0}\Gamma_n, \Gamma_0)$. One can prove that applying these operations to non-self-dual Wadge classes always produces non-self-dual Wadge classes. Moreover it follows from the proof of Theorem 8 in [**15**] that any non-self-dual Wadge class can be obtained from the trivial classes $\{\emptyset\}$ and $\{\check{\emptyset}\}$ by an iterative application of the operations **EXP** and **GSU**. For example $\mathbf{\Sigma}^0_1 = \mathbf{GSU}\big((\Gamma_n)_{n\geq 0}\big)$ with $\Gamma_0 = \{\emptyset\}$ and $\Gamma_n = \{\check{\emptyset}\}$ for all $n > 0$; and for all ξ, $\mathbf{\Sigma}^0_{1+\xi} = \mathbf{EXP}^{(\xi)}(\mathbf{\Sigma}^0_1)$.

To state properly this basic fact concerning the generation of Wadge classes one needs to introduce a notion of "*δ is a Wadge description of the class* Γ". One can find two such notions in [**13**] and [**15**] in connection with the respective generating operations considered in these works. Also it is not difficult to introduce a similar notion associated to our operations **EXP** and **GSU**; but for reasons which will be clarified later on, we need to achieve this with no reference to Wadge classes, so to define a notion of "*δ is a Wadge description*", which is of course a purely formal constraint.

<u>*The space of Wadge descriptions*</u>: Define inductively for $\mu, \lambda < \omega_1$:

$$\begin{cases} \mathbb{D}_0 = \{0,1\} \\ \mathbb{D}_{\mu+1} = \mathbb{D}_\mu \cup (\omega_1 \times \mathbb{D}_\mu) \cup (\mathbb{D}_\mu)^\omega \\ \mathbb{D}_\lambda = \bigcup_{\mu<\lambda} \mathbb{D}_\mu & \text{if } \lambda \text{ is limit} \end{cases}$$

The elements of $\mathbb{D}_0$ are said to be of level 0, and if $\delta \in \mathbb{D}_{\mu+1} \setminus \mathbb{D}_\mu$ we shall say that δ is a *Wadge description of level* μ. The space of all Wadge descriptions is then the set:

$$\mathbb{D} = \bigcup_{\mu<\omega_1} \mathbb{D}_\mu$$

We then define inductively for all $\delta \in \mathbb{D}$ a Wadge class $\mathbf{W}_\delta$ by:

$$\mathbf{W}_\delta = \begin{cases} \{\emptyset\} & \text{if } \delta \in \mathbb{D}_0 \text{ and } \delta = 0 \\ \{\check{\emptyset}\} & \text{if } \delta \in \mathbb{D}_0 \text{ and } \delta = 1 \\ \mathbf{EXP}^{(\xi)}(\mathbf{W}_\varepsilon) & \text{if } \delta \in \mathbb{D}_{\mu+1} \text{ and } \delta = (\xi, \varepsilon) \in \omega_1 \times \mathbb{D}_\mu \\ \mathbf{GSU}\big((\mathbf{W}_{\delta_n})_{n\geq 0}\big) & \text{if } \delta \in \mathbb{D}_{\mu+1} \text{ and } \delta = (\delta_n)_{n\geq 0} \in (\mathbb{D}_\mu)^\omega \end{cases}$$

One easily checks that each $\mathbf{W}_\delta$ is a non-self-dual Wadge class, and from the proof of Theorem 8 in [**15**], one gets the following result:

THEOREM 2.4.3. *Any non-self-dual Wadge class is of the form* $\mathbf{W}_\delta$ *for some* $\delta \in \mathbb{D}$.

Notice however that this parametrization is not one-to-one. For example for all $\xi \in \omega_1$ we have $\mathbf{W}_{(\xi,0)} = \mathbf{W}_0 = \{\emptyset\}$ and this phenomenon is exponentially amplified when dealing with descriptions of higher levels.

2.4.4. Extension of the operations EXP and GSU:. We already used each one of the symbols **EXP** and **GSU** to denote two formally different operations. We shall pursue and amplify this abuse of notation which as we shall see is very helpful for the intuition.

In the context of spaces $[R]$ of infinite branches the notion of distinction provides a natural way of extending the **EXP** operations. However recall that to collect mappings of arbitrary Baire class one has to restrict oneself to *regular resolutions of class* ξ (see 1.7.1). So given any tree R_0 we shall set:

$$\mathbf{EXP}^{(\xi)}(R_0) := \{R : R \text{ is a regular expansion of class } \xi \text{ of } R_0\}$$

which defines a *family* of subtrees of *Ext*.

We now introduce a similar extension of the operation **GSU** to trees. First if S is a tree and $(u_n)_{n>0}$ is a sequence in the domain of S, we shall say that $(u_n)_{n>0}$ is an <u>enumerated S-antichain</u> if the set $\{u_n;\ n > 0\}$ is an S-antichain, that is if:

$$(u_m\ S\ u_n \text{ or } u_n\ S\ u_m) \implies m = n$$

If $(R_n)_{n\geq 0}$ is a sequence of subtrees of a tree S, and $(u_n)_{n>0}$ is any enumerated S-antichain, the relation R defined by:

$$v\ R\ w \iff \begin{cases} v\ R_n\ w & \text{if } u_n\ S\ v \text{ for some (unique) } n > 0 \\ v\ R_0\ w & \text{if not} \end{cases}$$

is clearly a tree called the <u>orthogonal sum</u> of $(R_n)_{n\geq 0}$ over $(u_n)_{n>0}$; we shall also say more simply that R is an S-orthogonal sum of the sequence $(R_n)_{n\geq 0}$, and set:

$$\mathbf{GSU}^{(S)}((R_n)_{n\geq 0}) := \{R : R \text{ is an } S\text{-orthogonal sum of } (R_n)_{n\geq 0}\ \}$$

<u>*Extension to double-trees*</u>:

Let $(R_n^+, R_n^-)_{n\geq 0}$ be a sequence of double-trees and let for all n, R_n^* be the tree generated by (R_n^+, R_n^-). If R^+ , R^- , R^* are the orthogonal sums of $(R_n^+)_{n\geq 0}$, $(R_n^-)_{n\geq 0}$, $(R_n^*)_{n\geq 0}$ respectively, over the *same* enumerated S-antichain, then one easily checks that (R^+, R^-) is a double-tree generating the tree R^*; then we shall say that the double-tree (R^+, R^-) is an S-orthogonal sum of the sequence $(R_n^+, R_n^-)_{n\geq 0}$ of double-trees. Let $\mathbf{GSU}^{(S)}((R_n^+, R_n^-)_{n\geq 0})$ be:

$$\left\{(R^+, R^-) : (R^+, R^-) \text{ is an } S\text{-orthogonal sum of } (R_n^+, R_n^-)\right\}$$

2.4.5. Double-tree representation for Wadge classes: Recall that a double-tree (R^+, R^-) is said to be a double-tree representation in the tree R if $R^+ \cup R^- \subset R$ and the sets:

$$A^+ = \{\alpha \in [R] : \alpha \text{ contains an infinite } R^+\text{-branch}\ \}$$

and

$$A^- = \{\alpha \in [R] : \alpha \text{ contains an infinite } R^-\text{-branch}\ \}$$

form a partition of the space $[R]$. For simplicity we shall then say that the triple $(R^+, R^-; R)$ is a <u>*double-tree representation*</u>. This extends the notion introduced in Definition 2.1.4 which corresponds to the particular case where R is precisely the tree generated by the double-tree (R^+, R^-).

For any tree relation R let Δ_R denote the minimal subtree of R:

$$a\,\Delta_R\,b \iff a,b \in \mathrm{dom}(\Delta_R) \text{ and } (a=0 \text{ or } a=b)$$

For any Wadge description $\delta \in \mathbb{D}$, we define inductively a family $\mathscr{R}_\delta$ of double-tree representations as follows:

– if $\delta \in \mathbb{D}_0 = \{0,1\}$ then:

$\mathscr{R}_0 = \{(\Delta_R, R; R) : \ R \text{ tree relation } \}$ and $\mathscr{R}_1 = \{(R, \Delta_R; R); \ R \text{ tree relation } \}$

– if $\delta \in \mathbb{D}_{\mu+1}$ and $\delta = (\xi,\varepsilon) \in \omega_1 \times D_\mu$ then:

$$\mathscr{R}_\delta = \{(R^+, R^-; R) : R \in \mathbf{EXP}^{(\xi)}(R_0) \text{ and } (R^+, R^-; R_0) \in \mathscr{R}_\varepsilon\}$$

– if $\delta \in \mathbb{D}_{\mu+1}$ and $\delta = (\delta_n)_{n\geq 0} \in (\mathbb{D}_\mu)^\omega$ then:

$$\mathscr{R}_\delta = \{(R^+, R^-; R) : (R^+, R^-) \in \mathbf{GSU}^{(R)}\big((R_n^+, R_n^-)_{n\geq 0}\big) \text{ and } (R_n^+, R_n^-; R) \in \mathscr{R}_{\delta_n} \text{for all } n\}$$

If $(R^+, R^-; R) \in \mathscr{R}_\delta$ we shall say that $(R^+, R^-; R)$ is a Wadge double-tree representation of type δ, or that (R^+, R^-) is a Wadge double-tree representation in R of type δ.

The following result is an extension of Theorem 1.7.3 which corresponds to the particular case $\delta \in \mathbb{D}_2$ (since any Baire class has a description of level 2):

THEOREM 2.4.6. *Let R be a subtree of Ext and δ be a Wadge description. Then any Borel set $A \subset [R]$ in $\mathbf{W}_\delta$ admits a Wadge double-tree representation in R of type δ.*

PROOF. Suppose that $\delta \in \mathbb{D}_\mu$; we prove the result by induction on μ. The limit case (including the initial value $\mu = 0$) is trivial. So suppose that the result holds for all $\varepsilon \in \mu$ and consider $\delta \in \mathbb{D}_{\mu+1}$.

– If $\delta = (\xi,\varepsilon) \in \omega_1 \times \mathbb{D}_\mu$ then $A \in \mathbf{EXP}^{(\xi)}(B)$ for some $B \subset \omega^\omega$ in $\mathbf{W}_\varepsilon$. So there exists $\varphi : [R] \to \omega^\omega$ a Borel mapping of class ξ such that $A = \varphi^{-1}(B)$. For all $s \in \omega^{<\omega}$ set $A_s = \varphi^{-1}(N_s)$; then $(A_s)_{s\in\omega^{<\omega}}$ is a countable family of $\mathbf{\Sigma}^0_{1+\xi}$ subsets of $[R]$, and by Theorem 1.6.6 we can find a tree R_0 such that R is a regular expansion of class ξ of R_0 and if $\pi : [R_0] \to [R]$ is the canonical mapping then $\varphi \circ \pi$ is continuous. Hence $B_0 := (\varphi\circ\pi)^{-1}(B) \subset [R_0]$ is in $\mathbf{W}_\varepsilon$, and by the induction hypothesis B_0 admits a double-tree representation (R^+, R^-) in R_0 of type ε. But since $B_0 = \pi^{-1}(A)$ then clearly (R^+, R^-) is a double-tree representation of A in R of type δ.

– If $\delta = (\delta_n)_{n\geq 0} \in (\mathbb{D}_\mu)^\omega$ then $A \in \mathbf{GSU}((A_n)_{n\geq 0})$ with $A_n \subset [R]$ in $\mathbf{W}_{\delta_n}$ for all n. So there exists a sequence of pairwise disjoint basic clopen sets $(N_{s_n})_{n>0}$ such that

$$A = \Big(\bigcup_{n>0} A_n \cap N_{s_n}\Big) \cup \Big(A_0 \setminus \bigcup_{n>0} N_{s_n}\Big)$$

Applying the induction hypothesis, each A_n admits a double-tree representation (R_n^+, R_n^-) in R of type δ_n; and if R^+ and R^- are the orthogonal sums of $(R_n^+)_{n\geq 0}$ and $(R_n^-)_{n\geq 0}$ respectively, over the enumerated the R-antichain $(s_n)_{n>0}$, one easily checks that the double-tree (R^+, R^-) is a double-tree representation of A in R, which is obviously of type δ. □

CHAPTER 3

Two Applications of the Tree Representation

As explained in the Introduction, the representation of Borel sets developed in the previous chapters was elaborated for the needs of the proof of a result that we will tackle in Chapter 6. Incidentallly we discovered afterwards that this representation can be used first to give an alternative proof of some Hurewicz-type results due to Louveau – Saint Raymond, but also to solve Ostrovsky's problem discussed in the Introduction. Although none of these two results can be derived from the other, the proofs that we will give here will follow, in two totally different contexts, the same general scheme of construction. So we shall first present this construction in an "abstract" setting, and then apply it to the two mentioned particular situations.

3.1. Resolution of quasi-strategies

3.1.1. Games and quasi-strategies: In all the sequel E denotes a fixed set that we view as the set of all possible moves in some game with two players I and II.

We shall refer to any sequence $u \in \text{Seq}(E)$ as a position; if $|u|$ is odd (even) we shall say that u is a *I-position (II-position)* (in particular $u = \emptyset$ is a *II*-position). Also given *any* subset Λ of $\text{Seq}(E)$ we denote by Λ_I (Λ_{II}) the set of all I-positions (II-positions) in Λ.

We denote as usual by Ext the non strict extension relation on $\text{Seq}(E)$ and by Ext_I (Ext_{II}) its restriction to $\text{Seq}_I(E)$ ($\text{Seq}_{II}(E)$). So:

$$u\, Ext_I\, v \iff |u| \text{ and } |v| \text{ are odd and } u \preceq v$$

and similarly for Ext_{II}.

From now on we fix one of the players that we denote by J and denote by $\check{J}$ the opponent so $\{J, \check{J}\} = \{I, II\}$.

A game on E is just any sequential tree Γ on E (so $\Gamma \subset \text{Seq}(E)$). A quasi-strategy for *Player J* in the game Γ is a set $\Sigma \subset \Gamma$ satisfying:

(a) $\Sigma \subset \Gamma$ *and* Σ *is a sequential tree.*

(b) *For all* $u \in \Sigma_J$ *and* $u' \in \Gamma_{\check{J}}$ *if* $u \prec u'$ *with* $|u'| = |u|+1$ *there exists* $v \in \Sigma_J$ *such that* $u \prec u' \prec v$ *and* $|v| = |u'| + 1$

We shall say that $\Sigma \subset \text{Seq}(E)$ is a quasi-strategy for *Player J (without refering to any game)* if Σ is a sequential tree which is a quasi-strategy in (the game) Σ.

Let us emphasize here that in all this section game determinacy will be irrelevant and quasi-strategies will be used merely as a language to express some largeness notion. For this reason we avoided up to now to refer to any win condition. However we shall sometimes use the expression "*Player J wins the game* Γ" as a shortening for "*there exists a quasi-strategy for Player J in* Γ".

3.1.2. The games $\Gamma^{(R)}$: *From now on we fix a tree relation $R \subset Ext_J$ on* $\mathrm{Seq}_J(E)$.

Given any set $\Gamma \subset \mathrm{Seq}(E)$ we define the game $\Gamma^{(R)}$ *on* $\mathrm{Seq}(E)$ as follows. At each move in $\Gamma^{(R)}$ the players choose alternatively some $u_n \in \Gamma$ with the following rules:

(i) *If u_n is played by Player J then $u_n \in \Gamma_J$, $u_{n-2} \prec u_{n-1} \prec u_n$ and u_{n-2} is the R-predecessor of u_n.*

(ii) *If u_n is played by Player $\check{J}$ then $u_n \in \Gamma_{\check{J}}$, $u_{n-1} \prec u_n$ and $|u_n| = |u_{n-1}|+1$ where by convention $u_{-1} = u_{-2} = \emptyset$.*

Notice that $(u_0, u_1, \dots, u_n) \in \Gamma_J^{(R)}$ is entirely determined by $u_n \in \Gamma_J$. Thus formally speaking $\Gamma^{(R)}$ is a sequential tree *on* $\mathrm{Seq}(E)$ so $\Gamma^{(R)} \subset \mathrm{Seq}(\mathrm{Seq}(E))$. Of particular interest will be the case where $\Gamma \subset \mathrm{Seq}(E)$ is already a game on E; then the players' moves in $\Gamma^{(R)}$ are actually positions in the game Γ.

Notice also that if $\Sigma \subset \Gamma \subset \mathrm{Seq}(E)$ then $\Sigma^{(R)} \subset \Gamma^{(R)}$ are two sequential trees on $\mathrm{Seq}(E)$.

EXAMPLE 3.1.3. Let $J = II$ so $\mathrm{Seq}_J(E)$ is the set of all positions of even length, and let R be the tree relation on $\mathrm{Seq}_J(E)$ defined by

$$u \mathrel{R} v \iff u \preceq v \text{ and } |u| \equiv |v| \equiv 0 \pmod 4$$

In the game $\Gamma^{(R)}$ *Player I* chooses first some $u_0 \in \Gamma$ with $|u_0| = 1$, then *Player II* chooses $u_1 \in \Gamma$ with $u_0 \prec u_1$ and $|u_1| = 4$, then *Player I* chooses $u_2 \in \Gamma$ with $u_0 \prec u_1 \prec u_2$ and $|u_2| = 5$, then *Player II* chooses $u_3 \in \Gamma$ with $u_0 \prec u_1 \prec u_2 \prec u_3$ and $|u_3| = 8$, and so on ... Notice that u_0, u_1, u_2 are entirely determined by u_3, and a position $(u_0, u_1, \dots, u_n)$ in $\Gamma^{(R)}$ can be identified to the last move u_n.

In particular when Γ is a game on E then a run in $\Gamma^{(R)}$ can be viewed as a run in Γ in which we impose on *Player II* to play "long" *II*-positions in Γ (thus including some *I*-moves), while *Player I* is allowed from time to time to make a simple move as in Γ.

DEFINITION 3.1.4. Let $\Sigma \subset \Gamma \subset \mathrm{Seq}(E)$. We shall say that Σ is R-strategical for *Player J* in Γ if $\Sigma^{(R)}$ is a quasi-strategy for *Player J* in the game $\Gamma^{(R)}$.

If moreover $\Gamma = \Sigma$ then we shall simply say that Σ is R-strategical for *Player J*.

When $R = Ext_J$ the notion of R-strategical set coincides with the classical notion of quasi-strategy discussed in 3.1.1. But notice that in general an R-strategical set Σ is not necessarily a sequential tree. In fact this is the case if and only if the restriction of R to the set Σ_J coincides with Ext_J.

The following lemma is just the development of the previous definition.

LEMMA 3.1.5. *Σ is R-strategical for Player J in Γ if and only if:*

(a) *For all $v \in \Sigma_J$ if $u \prec u' \prec v$ with $|u'| = |u| + 1$ and $u \mathrel{R} v$ then $u' \in \Sigma_{\check{J}}$*
For all $u' \in \Sigma_{\check{J}}$ if $u \prec u' \prec v$ with $|u'| = |u| + 1$ then $u \in \Sigma_J$

(b) *For all* $u \in \Sigma_J$ *and* $u' \in \Gamma_{\check{J}}$ *if* $u \prec u'$ *with* $|u'| = |u| + 1$ *then there exists* $v \in \Sigma_J$ *such that* $u \prec u' \prec v$ *and* u *is the* R*-predecessor of* v.

3.1.6. The space $\mathscr{S}_J(R)$. We denote by $\mathscr{S}_J(R)$ the set of all R-strategical sets for *Player J*. Thus $\mathscr{S}_J(R)$ is a subset of the space $\mathscr{P}(\text{Seq}(E))$ which will be identified to the space $2^{\text{Seq}(E)}$ endowed with the product topology; and it is quite clear from the definitions that $\mathscr{S}_J(R)$ is a $\mathbf{\Pi}^0_2$ subset of the compact space $2^{\text{Seq}(E)}$, hence a Polish space.

The space $\mathscr{S}_J(R)$ should be viewed as a generalization of the space $[R]$ of all infinite R-branches. However while an element of $[R]$ is linearly ordered by R, an element $\Sigma \in \mathscr{S}_J(R)$ is a "tree-like" object. In fact R induces a tree relation on Σ_J but which, unlike for infinite branches, might be well founded (and even finite).

We recall (see Proposition 1.7.2) that if $R \subset S$ and S is a regular expansion of class ξ of R then any $\alpha \in [S]$ contains a unique $\hat{\alpha} \in [R]$ and the maping $\alpha \mapsto \hat{\alpha}$ (which is the inverse of the canonical mapping π) is of Baire class ξ. We shall now extend this fundamental result to the context of $\mathscr{S}_J(R)$ spaces .

We point out that for this generalization the inclusion relation between elements of $\mathscr{S}_J(R)$ and elements of $\mathscr{S}_J(S)$, is not fine enough and the right analog of "$\alpha' \in [R]$ *is contained in* $\alpha \in [S]$" will be "$\Sigma' \in \mathscr{S}_J(R)$ *is* R*-strategical for Player J in* $\Sigma \in \mathscr{S}_J(S)$"

THEOREM 3.1.7. *Let* $R \subset S \subset Ext_J$ *be two tree relations on the set* $\text{Seq}_J(E)$. *If* S *is a regular expansion of class* ξ *of* R *then there exists a mapping* $\Phi : \mathscr{S}_J(S) \to \mathscr{S}_J(R)$ *satisfying:*

1) For any $\Sigma \in \mathscr{S}_J(S)$, *the set* $\Phi(\Sigma)$ *is* R*-strategical for Player J in* Σ.

2) For all u *the set* $\{\Sigma : \ u \in \Phi(\Sigma)\}$ *is in* $\bigcup_{\eta<1+\xi} \mathbf{\Pi}^0_\eta$; *in particular* Φ *is of Baire class* ξ.

PROOF. In fact the mapping Φ will be obtained as the final step of a transfinite construction, and the verification of 1) and 2) will then be achieved by induction.

<u>Construction of Φ</u>: We fix a resolution family $\overrightarrow{R} = (R^{(\zeta)})_{\zeta \leq \xi^*}$ with $R^{(\xi^*)} = R$ and $R^{(0)} = S$. For any nonempty $u \in \text{Seq}(E)$ we denote by u^* the sequence obtained from u by deleting the last element, so $u^* \prec u$ and $|u^*| = |u| - 1$.

For any $\Sigma \in \mathscr{S}_J(S)$ we define inductively for $\eta + 1$ successor and λ limit $\leq \xi^*$:

$$\Sigma^{(0)}_{\cdot} = \Sigma$$

$$\begin{cases} \Sigma_J^{(\eta+1)} = \{u \in \Sigma_J^{(\eta)} \ : \ \ \forall v, w \in \Sigma_J^{(\eta)}, \text{ if } v\, R^{(\eta+1)} u \text{ and } v\, R^{(\eta)} w \text{ then } v R^{(\eta+1)} w\} \\[2ex] \Sigma_{\check{J}}^{(\eta+1)} = \{u \in \Sigma_{\check{J}}^{(\eta)} \ : \ \ u^* \in \Sigma_J^{(\eta+1)}\} \end{cases}$$

$$\Sigma^{(\lambda)}_{\cdot} = \bigcap_{\zeta<\lambda} \Sigma^{(\lambda)}_{\cdot}$$

Finally we define:

$$\Phi(\Sigma) = \Sigma^{(\xi^*)}$$

LEMMA 3.1.8. *For all* $\eta \leq \xi^*$ *the set* $\{\Sigma : \ u \in \Sigma^{(\eta)}\}$ *is* $\mathbf{\Pi}^0_\eta$.

PROOF. Observe that the sets $\{\Sigma :\ u \in \Sigma^{(0)}\}$ are $\mathbf{\Delta}^0_1$; moreover for all $\eta < \xi^*$ and all limit $\lambda \leq \xi^*$ we have:

$$u \in \Sigma_J^{(\eta+1)} \iff \begin{cases} u \in \Sigma_J^{(\eta)} \text{ and } \forall v, w \quad v \neg\, R^{(\eta+1)}\ u \text{ or } v \neg\, R^{(\eta)}\ w \\ \qquad \text{or } v\ R^{(\eta+1)}\ w \ \text{ or } v \notin \Sigma_J^{(\eta)} \text{ or } w \notin \Sigma_J^{(\eta)} \end{cases}$$

$$u \in \Sigma^{(\eta+1)} \iff u \in \Sigma_J^{(\eta+1)} \text{ or } u^* \in \Sigma_J^{(\eta+1)}$$

$$u \in \Sigma_J^{(\lambda)} \iff \forall \zeta < \lambda,\ u \in \Sigma_J^{(\zeta)}$$

and the lemma follows then by a straightforward induction. □

LEMMA 3.1.9. *If $\xi = \xi^*$ is limit and $\overrightarrow{R}$ is uniform then $\{\Sigma :\ u \in \Sigma^{(\xi)}\}$ is in $\bigcup_{\eta<\xi} \mathbf{\Pi}^0_\eta$*

PROOF. Set $n = h_R(u)$; then by assumption we can find $\eta < \xi$ such that:

(U) For all v, w if $\min(h_R(v)\,,\, h_R(w)) \leq n$ and $v\ R^{(\eta)}\ w$ then $v\ R\ w$

We shall prove next that

$$u \in \Sigma^{(\xi)} \iff u \in \Sigma^{(\eta+1)}$$

and the conclusion will then follow from Lemma 3.1.8.

So suppose that $u \in \Sigma^{(\eta+1)}$; and towards proving that $u \in \Sigma^{(\xi)}$ pick any $\zeta \in [\eta, \xi[$ and let v and w in Γ_J be such that: (i) $v\, R^{(\zeta+1)} u$; (ii) $v\, R^{(\zeta)} w$.

Since $h_R(u) = n$, it follows from (i) and (U) that $v\ R\ u$ hence $h_R(v) \leq h_R(u) = n$. Again since now $h_R(v) \leq n$, it follows from (ii) and (U) that $v\ R\ w$ hence $v\, R^{(\zeta+1)} w$.

This proves that $u \in \Sigma^{(\zeta+1)}$ for any ζ in $[\eta, \xi[$, hence that $u \in \Sigma^{(\xi)}$. □

LEMMA 3.1.10. *Φ is of Baire class ξ.*

PROOF. If ξ is successor then $1+\xi = \xi^*+1$ so by Lemma 3.1.8 the set $\{\Sigma :\ u \in \Sigma^{(\xi^*)}\}$ is in $\mathbf{\Pi}^0_{\xi^*} = \bigcup\{\mathbf{\Pi}^0_\eta;\ \eta < 1+\xi\}$; and by Lemma 3.1.9 the same holds in the limit case. In particular for all ξ the sets $\{\Sigma :\ u \in \Sigma^{(\xi^*)}\}$ are $\mathbf{\Delta}^0_{1+\xi}$ and it follows that their complements $\{\Sigma :\ u \notin \Sigma^{(\xi)}\}$ are also $\mathbf{\Delta}^0_{1+\xi}$; hence $\Phi^{-1}(\mathbf{\Sigma}^0_1) \subset \mathbf{\Sigma}^0_{1+\xi}$. □

LEMMA 3.1.11. *For all $\eta \leq \xi^*$, $\Sigma^{(\eta)} \in \mathscr{S}_J(R^{(\eta)})$.*

PROOF. We have to show that conditions *(a)* and *(b)* of Lemma 3.1.5 are satisfied (with $\Sigma = \Sigma^{(\eta)}$ and $\Gamma = \Sigma$).

The proof of condition *(a)* is straightforward by induction: the limit case is obvious and for the successor case notice that by its very definition $\Sigma^{(\eta+1)}$ is $R^{(\eta+1)}$-hereditary and condition *(a)* follows then from the equivalence: "$u \in \Sigma_{\bar{J}}^{(\eta+1)} \iff u^* \in \Sigma_J^{(\eta+1)}$".

So we only have to prove:

$(b\langle\eta\rangle)$: For all $u \in \Sigma_J^{(\eta)}$ and $u' \in \Sigma_{\bar{J}}$ if $u \prec u'$ with $|u'| = |u| + 1$ then there exists $v \in \Sigma_J$ such that $u \prec u' \prec v$ and u is the R-predecessor of v.

and for this we again proceed by induction.

For $\eta = 0$ we have $\Sigma^{(0)} = \Sigma$ and condition (b) is satisfied since by assumption $\Sigma \in \mathscr{S}_J(S)$ is R-strategical in Σ. For the inductive step we distinguish two cases:

The successor case: Assume $(b\langle\eta\rangle)$. To prove that $(b\langle\eta+1\rangle)$ holds we fix some $u \in \Sigma_J^{(\eta+1)}$ and $u' \in \Sigma_J$ such that $u \prec u'$ and $|u'| = |u| + 1$; we then have to find v in $\Sigma_J^{(\eta+1)}$ such that $u' \prec v$ and u is the $R^{(\eta+1)}$–predecessor of v. We argue by contradiction, assuming that such a v does not exist.

CLAIM 1. *There exists an infinite sequence $(v_n)_{n\geq 0}$ in $\Sigma_J^{(\eta)}$ satisfying for all n:*

(i) $v_n R^{(\eta)} v_{n+1}$

(ii) u is the $R^{(\eta+1)}$-predecessor of v_n.

Proof: We proceed by induction on n. Since $u \in \Sigma_J^{(\eta)}$ and $(b\langle\eta\rangle)$ holds, we can find $v_0 \in \Sigma_J^{(\eta)}$ such that $u' \prec v_0$ and u is the $R^{(\eta)}$-predecessor of v_0; hence $u\,R^{(\eta)} v_0$ and it follows then from the definition of $\Sigma_J^{(\eta+1)}$ that $u\,R^{(\eta+1)} v_0$, so a fortiori u is the $R^{(\eta+1)}$-predecessor of v_0.

Assume that $v_n \in \Sigma_J^{(\eta)}$ is already defined and satisfies (ii). Since $u \prec v_n$ and $u\,R^{(\eta+1)} v_n$, it follows from the contradiction assumption that $v_n \notin \Sigma_J^{(\eta+1)}$. But since all predecessors of v_n are already in $\Sigma_J^{(\eta+1)}$ (because $u \in \Sigma_J^{(\eta+1)}$) then we can find some $v_{n+1} \in \Sigma_J^{(\eta)}$ – that we choose of minimal length – such that $v_n\, R^{(\eta)} v_{n+1}$ but $v_n \neg R^{(\eta+1)} v_{n+1}$. It follows that $u\,R^{(\eta)} v_{n+1}$, and again since $u \in \Sigma_J^{(\eta+1)}$ then $u\,R^{(\eta+1)} v_{n+1}$. Finally let v^0_{n+1} denote the $R^{(\eta+1)}$-predecessor of v_{n+1}; we have to show that $v^0_{n+1} = u$.

First notice that $u\,R^{(\eta+1)} v^0_{n+1}\, R^{(\eta+1)} v_{n+1}$; and since $v_n\, R^{(\eta)} v_{n+1}$ then v_n and v^0_{n+1} are $R^{(\eta)}$-comparable: If v_n $R^{(\eta)}$-precedes v^0_{n+1} then $v_n\, R^{(\eta)} v^0_{n+1} R^{(\eta+1)} v_{n+1}$, but then $v_n \neg R^{(\eta+1)} v^0_{n+1}$ (for otherwise $v_n\, R^{(\eta+1)} v_{n+1}$); and this contradicts the minimal choice of v_{n+1}. Hence v^0_{n+1} $R^{(\eta)}$-precedes v_n, so $v^0_{n+1}\, R^{(\eta)} v_n\, R^{(\eta)} v_{n+1}$, but since $v^0_{n+1}\, R^{(\eta+1)} v_{n+1}$ then by the distinction of $R^{(\eta+1)}$ in $R^{(\eta)}$ we also have $v^0_{n+1}\, R^{(\eta+1)} v_n$. Finally we have $u\,R^{(\eta+1)} v^0_{n+1}\, R^{(\eta+1)} v_n$, and it follows then from the induction hypothesis of the Claim that $u = v^0_{n+1}$. ◇

We go back to the proof of the successor case. Since the sequence $\bar{v} := (v_n)_{n\geq 0}$ given by Claim 1 is an infinite $R^{(\eta)}$-chain, it is contained in some unique $R^{(\eta)}$-branch $\bar{u}$. On the other hand since S is an expansion of R then a fortiori $R^{(\eta)}$ is an expansion of $R^{(\eta+1)}$; hence $\bar{u}$ contains some infinite $R^{(\eta+1)}$-branch $\bar{w} := (w_n)_{n\geq 0}$. Since $\bar{v}$ and $\bar{w}$ are two compatible $R^{(\eta)}$-chains we can find indices m, n, p such that $v_0\, R^{(\eta)} w_m\, R^{(\eta)} v_n\, R^{(\eta)} w_p$ all relations being strict. But since $w_m\, R^{(\eta+1)} w_p$ then by distinction $w_m\, R^{(\eta+1)} v_n$, hence w_m is an $R^{(\eta+1)}$-predecessor of v_n strictly between u and v_n, and this contradicts (ii).

This contradiction finishes the proof of Lemma 3.1.11 in the successor case.

The limit case: The arguments are quite similar to the successor case, though slightly different; and we shall only indicate the modifications to bring to the previous case.

So suppose that $(b\langle\eta\rangle)$ holds for all $\eta < \lambda$ limit. Again to prove $(b\langle\lambda\rangle)$ we fix $u \in \Sigma_J^{(\lambda)}$ and $u' \in \Sigma_J^{(\lambda)}$ with $u \prec u'$ and $|u'| = |u| + 1$, and assume by contradiction that there is no $v \in \Sigma_J^{(\eta+1)}$ such that $u' \prec v$ and u is the $R^{(\eta+1)}$–predecessor of v.

CLAIM 2. *There exist an increasing sequence $(\eta_n)_{n\geq 0}$ converging to λ and an infinite sequence $(v_n)_{n\geq 0}$ satisfying for all n:*

(i) $v_n R^{(\eta_n)} v_{n+1}$

(ii) $v_n \in \Sigma_J^{(\eta_n)}$ *and* u *is the* $R^{(\eta_n+1)}$*-predecessor of* v_n

Proof: Again by induction on n. Since $u \in \Sigma_J^{(0)}$ we can find $v_0 \in \Sigma_J^{(0)}$ such that $u' \prec v_0$ and $|v_0| = |u'| + 1 = |u| + 2$. By assumption we cannot have $v_0 \in \Sigma_J^{(\lambda)}$ hence there exists some $\eta_0 < \lambda$ such that $v_0 \in \Sigma_J^{(\eta_0)} \setminus \Sigma_J^{(\eta_0+1)}$; then we necessarily have $u\, R^{(\eta_0)} v_0$ and since $u \in \Sigma_J^{(\lambda)} \subset \Sigma_J^{(\eta_0+1)}$ then $u\, R^{(\eta_0+1)} v_0$, so by the choice of $|v|$ obviously u is the $R^{(\eta_0+1)}$-predecessor of v_0

Assume that η_{n-1} and v_{n-1} are already defined and satisfy (ii). Since $u' \prec v_{n-1}$ and $u\, R^{(\eta_n+1)} v_{n-1}$, it follows from the assumption of the claim that $v_{n-1} \notin \Sigma_J^{(\lambda)}$, and as above there exists η_n such that $v_{n-1} \in \Sigma^{(\eta_n)} \setminus \Sigma^{(\eta_n+1)}$ and since $v_{n-1} \in \Sigma^{(\eta_{n-1})}$ then necessarily $\eta_n \geq \eta_{n-1}$. On the other hand $u \in \Sigma_J^{(\lambda)} \subset \Sigma_J^{(\eta_n+1)}$. Then replacing in the arguments of Claim 1 η by η_n we can find $v_n \in \Sigma_J^{(\eta_n)}$ such that u is the $R^{(\eta_n+1)}$-predecessor of v_n. ◇

To finish the proof notice that the S-chain $(v_n)_{n\geq 0}$ is contained in a unique infinite S-branch $\bar{u}$, and since S is an expansion of R, $\bar{u}$ contains an infinite R-chain $(w_n)_{n\geq 0}$. Fix m, n, p such that $u \prec w_m \prec v_n \prec w_p$. Then by Claim 2, $(v_k)_{k\geq n}$ and $(w_k)_{k\geq m}$ are two $R^{(\eta_n)}$-chains and replacing in the arguments of Claim 2 η by η_n one proves that u is not the $R^{(\eta_n+1)}$-predecessor of v_n contradicting thus condition (ii) of Claim 2. ◇

This finishes the proofs of Lemma 3.1.11 hence of Theorem 3.1.7. □

We finish this section by an extension of Theorem 3.1.7 to the context of Wadge classes. To state this result we need to introduce some terminology.

Given a class Γ; we shall say that a pair (A^+, A^-) of sets is a <u>$(\Gamma, \check{\Gamma})$ partition</u> of the set A if $A^+ \in \Gamma$, $A^- \in \check{\Gamma}$ and $\{A^+, A^-\}$ is a partition of A (we allow A^+ or A^- to be empty).

THEOREM 3.1.12. *Let* Γ *be a Wadge class of type* δ*, and* R *be tree relation on the set of all* J*-positions. If* (R^+, R^-) *is a Wadge double-tree representation in* R *of type* δ*, generating a tree* R^**, then there exists a Borel mapping* $\Phi : \mathscr{S}_J(R) \to \mathscr{S}_J(R^*)$ *such that:*

(1) *For all* $\Sigma \in \mathscr{S}_J(R)$, $\Phi(\Sigma)$ *is winning in* $\Sigma^{(R)}$.

(2) *For any position* u*, the set* $\mathscr{S}_u := \{\Sigma \in \mathscr{S}_J(R) : \; u \in \Phi(\Sigma)\}$ *is in* $\Delta(\Gamma)$ *and the pair:*

$$\left(\{\Sigma \in \mathscr{S}_u : \textit{Player } J \textit{ wins } \Phi(\Sigma)^{(R^+)}\},\ \{\Sigma \in \mathscr{S}_u : \textit{Player } J \textit{ wins } \Phi(\Sigma)^{(R^-)}\}\right)$$

is a $(\Gamma, \check{\Gamma})$ *partition of* $\mathscr{S}_u$.

PROOF. Suppose that $\delta \in \mathbb{D}_\mu$; the proof which is by induction on μ follows a scheme similar to the proof of Theorem 2.4.6, and we only sketch the main steps.

The limit case is also trivial here, and we can restrict ourselves to the successor case. So suppose that the result holds for all $\varepsilon \in \mathbb{D}_\mu$ and consider $\delta \in D_{\mu+1}$.

– If $\delta = (\xi, \varepsilon) \in \omega_1 \times \mathbb{D}_\mu$ then R is a regular expansion of class ξ of some tree R_0 such that (R^+, R^-, R_0) is a Wadge double-tree representation of type ε. The induction hypothesis provides a Borel mapping $\Phi_0 : \mathscr{S}_J(R_0) \to \mathscr{S}_J(R^*)$ and

Theorem 3.1.7 provides a Borel mapping $\Phi_1 : \mathscr{S}_J(R) \to \mathscr{S}_J(R_0)$. Define then the mapping $\Phi : \mathscr{S}_J(R) \to \mathscr{S}_J(R^*)$ by $\Phi := \Phi_0 \circ \Phi_1$

– If $\delta = (\delta_n)_{n\geq 0} \in (\mathbb{D}_\mu)^\omega$ then (R^+, R^-) is the orthogonal sum of some sequence $(R_n^+, R_n^-)_{n\geq 0}$ of double-trees over some enumerated the R-antichain $(u_n)_{n>0}$, such that $(R_n^+, R_n^-; R) \in \mathscr{R}_{\delta_n}$ for all n. The induction hypothesis provides for all n a Borel mapping $\Phi_n : \mathscr{S}_J(R) \to \mathscr{S}_J(R_n^*)$ (where R_n^* is the tree generated by (R_n^+, R_n^-)). Define then the mapping $\Phi : \mathscr{S}_J(R) \to \mathscr{S}_J(R^*)$ by:

$$u \in \Phi(\Sigma) \iff \begin{cases} u \in \Phi_n(\Sigma) & \text{if } u_n \prec u \\ u \in \Phi_0(\Sigma) & \text{if not} \end{cases}$$

for all $\Sigma \in \mathscr{S}_J(R)$.

We leave to the reader to check that in each case the corresponding mapping Φ possesses the desired properties. □

3.2. Hurewicz type results

Our first application of Theorem 3.1.7 is a new proof of the following result due to Louveau – Saint Raymond ([**14**]) extending a classical result of Hurewicz. This proof provides a nice and simple illustration of Theorem 3.1.7. The arguments will make use of double-trees merely as an intuitive language, but the technical results on the double-tree representation developed in the previous chapter are not needed here.

(Louveau, Saint Raymond) Let A_0 and A_1 be two disjoint $\mathbf{\Sigma}_1^1$ sets in 2^ω. Let B_0 be a $\mathbf{\Pi}_{1+\xi}^0$ subset of 2^ω, and set $B_1 = 2^\omega \setminus B_0$. Then one of the following holds:

(1) : A_0 is separated from A_1 by a $\mathbf{\Sigma}_{1+\xi}^0$ set.

(2) : There exists $\varphi : 2^\omega \to 2^\omega$ continuous such that $\varphi(B_0) \subset A_0$ and $\varphi(B_1) \subset A_1$.

Proof. We identify as usual $(2\times\omega)^{<\omega}$ with $\bigcup_n (2^n \times \omega^n)$ and $(2\times\omega)^\omega$ with $2^\omega \times \omega^\omega$, and fix for $i = 0, 1$, a sequential tree T_i on $2\times\omega$ such that A_i is the projection of $\lceil T_i \rceil$ on the first factor.

We shall say that $(t,a) \in \mathrm{Seq}(2)\times\mathrm{Seq}(\omega)$ is <u>compatible</u> with T_i if $(t_{|k}, a_{|k}) \in T_i$ with $k = \min(|t|, |a|)$

We fix $(F, R, \overrightarrow{R})$ a regular $\mathbf{\Pi}_1^0$-representation of class ξ for the $\mathbf{\Pi}_{1+\xi}^0$ set $B_0 \subset 2^\omega \approx [Ext]$ (see 1.7.4): we recall that $\overrightarrow{R} = (R^{(\eta)})_{\eta\leq\xi^*}$ is a regular expansion of class ξ for R in *Ext* and F is a $\mathbf{\Pi}_1^0$ subset of $[R]$. Finally let (R^+, R^-) be the canonical double-tree representation of F (see Example 2.1.1) defined by:

$$\begin{cases} s\, R^+\, t \iff s\, R\, t \text{ and } N_s \cap F \neq \emptyset \\ a\, R^-\, b \iff s\, R\, t \text{ and } (N_s \cap F = \emptyset \text{ or } s = 0) \end{cases}$$

where N_s denotes the basic open set N_s^R of $[R]$.

<u>The game G</u>: We now consider the *open* game G in which at their respective $(n+1)^{\text{th}}$ move, *Player I* chooses $s_n \in 2^n$ and *Player II* chooses $(t_n, a_n) \in 2^n \times \omega^{\leq n}$ such that for all $0 < m < n$:

(i) $s_m \prec s_n$ and $t_m \prec t_n$

(ii) (t_n, a_{2n}) is compatible with T_0 and (t_n, a_{2n+1}) is compatible with T_1

(iii) If $s_m\, R^+\, s_n$ then $a_{2m} \prec a_{2n}$

(iv) If $s_m\ R^-\ s_n$ then $a_{2m+1} \prec a_{2n+1}$

We shall say that $y = \bigcup_n s_n \in 2^\omega$ is the *real constructed by Player I* in the run, and $z = \bigcup_n t_n \in 2^\omega$ is the *real constructed by Player II* in this run.

Since the game G is determined, Theorem 3.2 will follow from next two lemmas.

LEMMA 3.2.1. *If Player II wins the game G then (2) holds.*

PROOF. Fix a winning strategy τ for *Player II* in G. For all $y \in 2^\omega$ let $\varphi(y)$ be the real constructed by *Player II* in the unique infinite run $\rho(y)$ compatible with τ in which *Player I* constructs y. Then the mapping $\varphi : 2^\omega \to 2^\omega$ thus defined is clearly continuous, and even Lipschitz.

We now prove that $\varphi(B_i) \subset A_i$. So fix $y \in 2^\omega$ that we identify to $(y_{|n})_{n\geq 0} \in [Ext]$; let (t_n, a_n) be the $(n+1)^{\text{th}}$ response by τ in the run $\rho(y)$, and let $\hat{y} \in [R]$ be the unique infinite R-branch contained in y (identified with the Ext-branch $(y_{|n})_{n\geq 0}$):

– If $y \in B_0$ let $= (y_{|m_j})_{j\geq 0}$ be the infinite R^+-branch contained in $\hat{y}$. Then by rule (iii) $a_{2m_j} \prec a_{2m_k}$ for all $j < k$, hence $\bigcup_{j\geq 0}(t_{m_j}, a_{2m_j}) \approx (z, \alpha)$ can be identified to an element of $\lceil T_0 \rceil$, so $\varphi(y) = z \in A_0$.

– If $y \in B_1$ let $= (y_{|m_j})_{j\geq 0}$ be the infinite R^--branch contained in $\hat{y}$. Then by rule (iv) $a_{2m_j+1} \prec a_{2m_k+1}$ for all $j < k$, hence $\bigcup_{j\geq j_0}(t_{m_j}, a_{2m_j+1}) \approx (z, \alpha')$ can be identified to an element of $\lceil T_1 \rceil$, so $\varphi(y) = z \in A_1$. □

LEMMA 3.2.2. *If Player I wins the game G then (1) holds.*

PROOF. Fix a winning quasi-strategy Σ for *Player I* in G.

For any $z \in 2^\omega$ let $\Sigma(z)$ denote the set of all positions in Σ compatible with z in the sense that all moves of *Player II* in $\Sigma(z)$ are of the form $(z_{|n}, a_n)$. We view each $\Sigma(z)$ as a quasi-strategy for a game in which *Player I* chooses $s_n \in 2^n$ and *Player II* chooses $a_n \in \omega^{\leq n}$ such that the $(s_n, z_{|n}, a_n)$ satisfy the rules of the game G. The rules of the game G induce conditions on $\Sigma(z)$ to which we still refer by (i), (ii), (iii), (iv). Notice that although the game G is just open, the game $\Sigma(z)$ is clopen.

We shall view any position in the game $\Sigma(z)$ as an element of the set Λ of all finite sequences with values alternatively in: $2^{<\omega}$ (for even indices) and $\omega^{<\omega}$ (for odd indices). As usual we denote by Λ_I (Λ_{II}) the set of elements of Λ of odd (even) length, and for any z we set $\Sigma_I(z) := \Sigma(z) \cap \Lambda_I(z)$.

If $u = (s_0, a_0, \ldots\ldots, s_{n-1}, a_{n-1}, s_n) \in \Lambda_I$, we shall say that u is of type 0 (type 1) if n is even (odd). We denote by Λ_0 (Λ_1) the set of all $u \in \Lambda_I$ of type 0 (of type 1), and for $j \in \{0, 1\}$ we also set $\Sigma_j(z) := \Sigma(z) \cap \Lambda_j(z)$; so $\Sigma_I(z) = \Sigma_0(z) \cup \Sigma_1(z)$. Notice that by rule (ii), if $u \in \Sigma_j(z)$ and $a_n \in \omega^{\leq n}$ is a legal move for *Player II* in the game $\Sigma(z)$ then $a_n \in T_j(z)$. Finally if $u \in \Lambda_I$ is as above and p is the integer part of $\dfrac{n}{2}$ we set:

$$\chi(u) := s_p$$

then rules (iii) and (iv) of the game G can be transcribed as follows: for any $u, v \in \Sigma_I(z)$ and any $a, b \in 2^{<\omega}$ such that $u^\frown\langle a\rangle$ and $v^\frown\langle b\rangle$ are licit II-positions in $\Sigma(z)$ with $u^\frown\langle a\rangle \prec v^\frown\langle b\rangle$, then:

$$\begin{cases} (u, v \text{ are of type 0 and } \chi(u)\ R^+\ \chi(v)) \implies a \prec b \\ (u, v \text{ are of type 1 and } \chi(u)\ R^-\ \chi(v)) \implies a \prec b \end{cases}$$

Consider now for all $\eta \leq \xi^*$ the tree relation $\tilde{R}^{(\eta)}$ on Λ_I defined by:

$$u\,\tilde{R}^{(\eta)}v \iff u \preceq v \text{ and } \chi(u)\,R^{(\eta)}\chi(v)$$

Since $(R^{(\eta)})_{\eta\leq\xi^*}$ is a regular resolution for $R = R^{(\xi^*)}$ in $Ext_I = R^{(0)}$ then obviously $(\tilde{R}^{(\eta)})_{\eta\leq\xi^*}$ is a regular resolution for $\tilde{R} := \tilde{R}^{(\xi^*)}$ in $Ext_I = \tilde{R}^{(0)}$.

Let $\Phi : \mathscr{S}_I(Ext_I) \to \mathscr{S}_I(\tilde{R})$ be the Borel mapping given by Theorem 3.1.7 applied with $J = I$. For all $z \in 2^\omega$ set:

$$\Sigma^*(z) = \Phi(\Sigma(z))$$

and consider the set:

$$C_0 = \{z \in 2^\omega : \exists\, u \in \Sigma_1^*(z) : N_{\chi(u)} \cap F = \emptyset\}$$

then by condition 2) of Theorem 3.1.7, the set C_0 is $\mathbf{\Sigma}^0_{1+\xi}$ and we shall now prove that it separates A_0 from A_1, which will establish condition (1) of Theorem 3.2.

Claim 1. $A_0 \subset C_0$.

Proof: We argue by contradiction: suppose that there exists some $z \in A_0 \setminus C_0$ and fix α such that $(z, \alpha) \in \lceil T_1 \rceil$. Notice that since $z \notin C_0$ then for any $u \in \Sigma_0^*(z)$, $N_{\chi(u)} \cap F \neq \emptyset$, hence the R^--predecessor of $\chi(u)$ is the empty sequence $\emptyset$, consequently $u^\frown\langle\emptyset\rangle$ is a legal II-position in the game $\Sigma(z)$. Moreover for all $u, v \in \Sigma_0^*(z)$ such that $u \prec v$ we necessarily have $\chi(u)\ R^+\ \chi(v)$.

Condition 1) of Theorem 3.1.7 ensures that $\big(\Sigma(z)\big)^{(\tilde{R})}$ is a quasi-strategy for *Player I* in the game $\big(\Sigma^*(z)\big)^{(\tilde{R})}$ (see 3.1.2). Consider then in this game the run in which *Player I* follows the quasi-strategy $\big(\Sigma^*(z)\big)^{(\tilde{R})}$, and to any move $v \in \Sigma_I^*(z)$ made by *Player I*:

– if $v \in \Sigma_0^*(z)$ then *Player II* answers a suitable beginning of α.

– if $v \in \Sigma_1^*(z)$ then *Player II* answers $\emptyset$

By the observations above such a run is well defined (that is all *Player II*'s moves are legal). It is also clearly infinite, which is impossible, since the quasi-strategy $\big(\Sigma^*(z)\big)^{(\tilde{R})}$ is winning for *Player I* in $\big(\Sigma(z)\big)^{(\tilde{R})}$. ◇

This finishes the proof of Lemma 3.2.2. □

Claim 2. $C_0 \cap A_1 = \emptyset$.

Proof: The argument is similar and goes also by contradiction: suppose that there exists some $z \in C_0 \cap A_1$ and fix α such that $(z, \alpha) \in \lceil T_1 \rceil$. Since $z \in C_0$ we can also fix $u_0 \in \Sigma_1^*(z)$ of minimal length such that $N_{\chi(u_0)} \cap F = \emptyset$; for simplicity we shall assume that $\chi(u_0) \neq \emptyset$. Let $\bar{u}$ denote the unique position in $\big(\Sigma^*(z)\big)^{(\tilde{R})}$ ending by u_0. Since $\chi(u_0) \neq \emptyset$ and u_0 is of type 1 then $\bar{u}$ is necessarily of the form:

$$\bar{u} = (\ldots\ldots, a_0^*, u_0^*, b_0, v_0, a_0, u_0)$$

where u_0^*, v_0, u_0 (a_0^*, b, a_0) are the three last moves made by *Player I* (*Player II*) in $\bar{u}$. Notice that $v_0 \in \Sigma_0^*(z)$ is *the* $\tilde{R}$-predecessor of u_0, and $u_0^* \in \Sigma_1^*(z)$ is the largest $\tilde{R}$-predecessor of u_0 in $\Sigma_1^*(z)$, hence $a_0^*, a_0 \in T_0(z)$ and $b \in T_1(z)$.

Since u_0 is minimal, then $N_{\chi(u_0^*)} \cap F \neq \emptyset$, hence $\chi(u_0^*)\ R^+\ \chi(u_0)$ and a fortiori $\chi(u_0^*)$ is the R^+-predecessor of $\chi(u_0)$. Moreover by rule (iii) of the game, $a_0^* \prec a_0$. It follows then from the definition of (R^+, R^-) that if v and w are any I-positions in $\Sigma(z)$ such that $u_0 \tilde{R} v \tilde{R} w$, then necessarily $\chi(u_0^*) R^- \chi(v) R^- \chi(w)$, and $\chi(u_0^*)$ is the R^+-predecessor of both $\chi(v)$ and $\chi(w)$; in particular if $v \in \Sigma_0(z)$ then $v^\frown\langle a_0\rangle$ respects rule (iii) and is a legal II-position in $\Sigma(z)$.

Starting from $\bar{u}$ consider then in the game $\left(\Sigma(z)\right)^{(\tilde{R})}$ the run in which *Player I* follows the quasi-strategy $\left(\Sigma^*(z)\right)^{(\tilde{R})}$, and to any move $v \in \Sigma_I^*(z)$ made by *Player I*:

– if $v \in \Sigma_0^*(z)$ then *Player II* answers a_0

– if $v \in \Sigma_1^*(z)$ then *Player II* answers a suitable beginning of α.

Again such a run is well defined and infinite, which is impossible. ◇

This finishes the proof of Lemma 3.2.2 hence of Theorem 3.2. □

REMARKS 3.2.3. *a)* With the same arguments, one can give an effective version of Theorem 3.2 as in [**14**]. Also, assuming that $\aleph_\xi^L$ is countable, the proof is easily adaptable to the case where A_0 and A_1 are Σ_2^1.

b) The original proof of the previous result in [**14**] is also based on some game $\tilde{G}$. However unlike the game G considered in the present proof, which is in an obvious sense quite natural, the game $\tilde{G}$ is quite complicated and defined transfinitely. Moreover the proof in [**14**] is achieved in two steps: first for a particular family of Borel sets which are explicitly constructed, from which the general case is then derived. A more significant difference is that the separating set C_0 constructed in the above proof has some structural properties. For example one can show that

$$C_0 = \{z \in 2^\omega : \textit{Player I wins the game } \Phi(\Sigma(z))^{(R^-)}\}$$

This observation will be more explicit in the proof of the following result, also due to Louveau and Saint Raymond ([**15**]), which extends Theorem 3.2 to Wadge classes.

THEOREM 3.2.4. (Louveau, Saint Raymond) *Let A_0 and A_1 be two $\mathbf{\Sigma}_1^1$ sets in 2^ω. Let Γ be a Wadge non-self-dual class, let B_0 be a Borel subset of 2^ω in Γ, and set $B_1 = 2^\omega \setminus B_0$. Then one of the following alternatives holds:*

(1): A_0 is separated from A_1 by a set in $\check{\Gamma}$.

(2): There exists $\varphi : 2^\omega \to 2^\omega$ continuous such that $\varphi(B_0) \subset A_0$ and $\varphi(B_1) \subset A_1$.

PROOF. The proof follows the lines of the proof of Theorem 3.2, and we shall only indicate the additional needed arguments.

First observe that by Theorem 2.4.3, $\Gamma = \mathbf{W}_\delta$ for some Wadge description δ. Fix for $i = 0$, 1, a sequential T_i tree on $2 \times \omega$ such that A_i is the projection of $\lceil T_i \rceil$ on the first factor. Then applying Theorem 2.4.6 to $B_0 \subset 2^\omega \approx [Ext]$ fix a Wadge double-tree representation (R^+, R^-) of B_0 in Ext of type δ. Finally consider the game $G := G_{[R^+,R^-,T_0,T_1]}$ introduced in the proof of Theorem 3.2.

Notice that the arguments of Lemma 3.2.1 are still valid, and if *Player II* wins G then alternative (2) holds. Hence to prove Theorem 3.2.4 it is enough to prove that in this new context, if *Player I* wins G then alternative (1) holds.

So fix a winning strategy Σ for *Player I* in G, and let for all $z \in 2^\omega$, $\Sigma(z)$ denote the set of all positions in Σ compatible with z in the sense of the proof of

Theorem 3.2. Let Φ be the mapping given by Theorem 3.1.12 and consider the sets:

$$C^+ := \{z \in 2^\omega : \textit{Player I wins the game } \Phi(\Sigma(z))^{(R^+)}\}$$

$$C^- := \{z \in 2^\omega : \textit{Player I wins the game } \Phi(\Sigma(z))^{(R^-)}\}$$

By Theorem 3.1.12 (applied to $u = \emptyset$), the pair (C^+, C^-) forms a $(\Gamma, \check{\Gamma})$ partition of 2^ω, and we now prove that C^- separates A_0 from A_1.

CLAIM . *$C^+ \cap A_0 = \emptyset$ and $C^- \cap A_1 = \emptyset$.*

Proof: Suppose for example that there exists $z \in C^+ \cap A_0$ and fix $\alpha \in \omega^\omega$ such that $(z, \alpha) \in \lceil T_0 \rceil$. Consider then a run in the game $(\Sigma(z))^{(R^+)}$ in which *Player I* follows the quasi-strategy $\Phi(\Sigma(z))^{(R^+)}$ and *Player II* plays at each move a suitable beginning of α. In this run both players can play for ever, and this impossible since by part 1) of Theorem 3.1.12, the quasi-strategy $\Phi(\Sigma(z))^{(R^+)}$ is winning in the game $(\Sigma(z))^{(R^+)}$.

The same argument shows that $C^- \cap A_1 = \emptyset$. ◇

This finishes the proof of Theorem 3.2.4. □

REMARK 3.2.5. In the particular case $\Gamma = \mathbf{\Pi}^0_{1+\xi}$, the set C^- provided by the previous proof is essentially isomorphic to the set C_0 constructed in the proof of Theorem 3.2. In fact since the Baire class $\mathbf{\Pi}^0_\xi$ is obtained from the trivial class $\{\emptyset\}$ by applying successively the operation **GSU** and then $\mathbf{EXP}^{(\xi)}$, it has a Wadge description $\delta \in \mathbb{D}_2$. Consequently, unlike C_0 which was defined explicitly, the set C^- is obtained here in two elementary steps.

3.3. A Borel separation result

We now come to the second application of Theorem 3.1.7. This result constitutes the main step for the solution of Ostrovsky's Problem, which we will develop in next chapter.

Lipschitz mapping: If X is an arbitrary subset of a sequential space A^ω, let T_X denote the canonical sequential tree defining its closure $\overline{X}$ in A^ω, so

$$T_X := \{s \in \mathrm{Seq}(A) : N_s \cap X \neq \emptyset\}$$

We shall then say that the mapping $\varphi : X \longrightarrow B^\omega$ is *Lipschitz* if there exists a (uniquely determined) mapping $\phi : T_X \longrightarrow \mathrm{Seq}(B)$ satisfying:

(i) $|\phi(s)| = |s|$; (ii) if $s \prec t$ then $\phi(s) \prec \phi(t)$; (iii) $\varphi(x) = \bigcup_n \phi(x_{|n})$ for all $x \in X$.

The mapping ϕ will be called the *Lipschitz extension mapping* of φ.

Obviously a Lipschitz mapping $\varphi : X \longrightarrow B^\omega$ admits a unique Lipschitz extension to $\overline{X}$.

THEOREM 3.3.1. *Let $\pi : 2^\omega \times 2^\omega \longrightarrow 2^\omega$ denote the canonical projection on the first factor, $X \subset 2^\omega \times 2^\omega$ be $\mathbf{\Pi}^0_{1+\xi}$ and $Y \subset 2^\omega$ be $\mathbf{\Sigma}^1_1$. If any compact subset of Y admits a Lipschitz lifting in X then there exists a $\mathbf{\Pi}^0_{1+\xi}$ set B such that $Y \subset B \subset \pi(X)$.*

In particular if any compact subset of $\pi(X)$ admits a Lipschitz lifting in X then $\pi(X)$ is Borel of class $\mathbf{\Pi}^0_{1+\xi}$.

PROOF. We fix a sequential tree $T \subset \mathrm{Seq}(2 \times \omega)$ such that Y is the projection of $\lceil T \rceil \subset (2 \times \omega)^\omega \approx 2^\omega \times \omega^\omega$ on the first factor.

Again we shall say that $(t, a) \in \mathrm{Seq}(2) \times \mathrm{Seq}(\omega)$ is *compatible* with T if $(t_{|k}, a_{|k}) \in T$ with $k = \min(|t|, |a|)$

We identify $2^\omega \times 2^\omega$ with the space $[S]$ where S is the tree relation on $E = \bigcup_n 2^n \times 2^n$ defined by:

$$(s, t)\ S\ (s', t') \iff s \preceq s' \text{ and } t \preceq t'$$

and fix $(F, R, \overrightarrow{R})$ a regular $\mathbf{\Pi}^0_1$-representation of class ξ for the $\mathbf{\Pi}^0_{1+\xi}$ subset X of $[S]$. So $\overrightarrow{R} = (R^{(\eta)})_{\eta \leq \xi^*}$ is a regular expansion of class ξ for R in *Ext* and F is a $\mathbf{\Pi}^0_1$ subset of $[R]$.

Finally for any $(s, t) \in E$ we denote simply by $N_{(s,t)}$ the basic open set $N^R_{(s,t)}$ of $[R]$.

<u>The game G</u>: We now consider the *closed* game G in which:

– at his $(n+1)^{\text{th}}$ move *Player I* chooses $(s_n, a_n) \in 2^n \times \omega^{\leq n}$ such that:
  (i) (s_n, a_n) is compatible with T
  (ii) $s_m \prec s_n$ for all $m < n$

– at his $(n+1)^{\text{th}}$ move *Player II* chooses $t_n \in 2^n$ such that:
  (iii) $t_m \prec t_n$ for all $m < n$

<u>*Win condition*</u>: Thus in an infinite run in G *Player I* constructs two infinite sequences $y = \bigcup_n s_n \in 2^\omega$ and $\bar{a} = (a_n)_{n \in \omega} \in (\mathrm{Seq}(\omega))^\omega$ (notice that a priori a_n does not extend a_{n-1}), and *Player II* constructs an infinite sequence $z = \bigcup_n t_n \in 2^\omega$.

Player I wins the run if the following two conditions are fulfilled:

(1) $(y, z) \notin X$
(2) if $\widehat{(y, z)} := (y_{|n_j}, z_{|n_j})_{j \in \omega}$ is the unique R-branch contained in (y, z) then:

$$\exists k,\ \forall j > i > k : \quad a_{n_i} \prec a_{n_j}$$

We recall that, by the identification $2^\omega \times 2^\omega \approx [S]$, we view (y, z) as an infinite S-branch.

The win condition of G being clearly Borel, this game is determined and Theorem 3.3.1 will follow directly from the next two lemmas.

LEMMA 3.3.2. *If Player I wins the game G then there exists a compact subset of Y with no Lipschitz lifting in X.*

PROOF. A winning strategy σ for *Player I* in G determines a Lipschitz mapping f from 2^ω to $2^\omega \times (\mathrm{Seq}(\omega))^\omega$ such that in any infinite run compatible with σ, if *Player II* constructs $z \in 2^\omega$ then *Player I* constructs $f(z) = (y, \bar{a}) \in 2^\omega \times (\mathrm{Seq}(\omega))^\omega$. Notice that by the win condition (1) we necessarily have $(y, z) \notin X$; moreover:

CLAIM . $y \in Y$

<u>*Proof*</u>: Let $(y_{|n_j}, z_{|m_j})_{j \geq 0}$ be the unique R-branch contained in (y, z). By the win condition (2) there exists some index k such that $a_{m_i} \prec a_{m_j}$ for all $j > i \geq k$, hence $\bigcup_{j \geq k}(s_{m_j}, a_{m_j})$ can be identified to an element (y, α) of $\lceil T \rceil$, and by the choice of T this ensures that $y \in Y$. ◇

Consider then the range space K of f: it follows from the Claim and the continuity of f, that K is a compact subset of Y. If $\varphi : K \to 2^\omega$ is Lipschitz with graph contained in X then there exists an infinite run in which *Player I* follows σ to construct some $(y, \bar{a}) \in 2^\omega \times (\mathrm{Seq}(\omega))^\omega$ and *Player II* plays at his $(n+1)^{\text{th}}$ move $t_n = \phi(s_n)$ thus constructs an *infinite* sequence $z \in 2^\omega$ such that $z = \varphi(y)$; hence $(y, z) \in X$ and *Player II* wins the run, which is impossible. This proves that K does not admit any Lipschitz lifting in X. □

LEMMA 3.3.3. *If Player II wins the game G then there exists a* $\mathbf{\Pi}^0_{1+\xi}$ *set* B *such that* $Y \subset B \subset \pi(X)$.

PROOF. Fix a winning strategy τ for *Player II* in G. For any $y \in 2^\omega$ let $\Sigma(y)$ denote the set of all positions in the game G compatible with τ and y in the sense that all moves of *Player I* are of the form $(y_{|n}, a_n)$.

Let Λ denote the set of all II-positions in the game G. An element $u \in \Lambda$ can be identified with a sequence $((a_0, s_0, t_0), \ldots\ldots, (a_n, s_n, t_n)) \in \mathrm{Seq}(\mathrm{Seq}(\omega) \times \mathrm{Seq}(2) \times \mathrm{Seq}(2))$ with $|s_n| = |t_n|$, and we shall set:

$$\chi(u) = (s_n, t_n)$$

For all $\eta \leq \xi$ if we define the relation $\tilde{R}^{(\eta)}$ on Λ by:

$$u\, \tilde{R}^{(\eta)} v \iff u \preceq v \text{ and } \chi(u)\, R^{(\eta)} \chi(v)$$

Since $(R^{(\eta)})_{\eta \leq \xi^*}$ is a regular resolution family of class ξ for $R = R^{(\xi^*)}$ in $S = R^{(0)}$ then obviously $(\tilde{R}^{(\eta)})_{\eta \leq \xi^*}$ is also a regular resolution family of class ξ for $\tilde{R} := \tilde{R}^{(\xi^*)}$ in $\tilde{R}^{(0)} = S$.

For all $y \in 2^\omega$ we can view the set $\Sigma_{II}(y)$ of all II-positions in the game $\Sigma(y)$ as a subset of Λ, on which the family $(\tilde{R}^{(\eta)})_{\eta \leq \xi^*}$ induces a regular resolution of class ξ for $\tilde{R} := \tilde{R}^{(\xi^*)}$ in $\tilde{S} := \tilde{R}^{(0)}$. Let Φ be the associated Borel mapping given by Theorem 3.1.7 and set

$$\Sigma^*(y) := \Phi(\Sigma(y))$$

Since for all $u \in \Lambda$ the set $\{y : u \in \Sigma^*(y)\}$ is $\mathbf{\Pi}^0_\eta$ for some $\eta < 1 + \xi$, then the set

$$B = \{y \in 2^\omega : \forall u \in \Sigma^*_{II}(y),\ N_{\chi(u)} \cap F \neq \emptyset\}$$

is $\mathbf{\Pi}^0_{1+\xi}$; and we shall now show that $Y \subset B \subset \pi(X)$.

We first prove that $B \subset \pi(X)$. Recall that for all y, $\Sigma^*(y)$ is a quasi-strategy for *Player II* in the game $\Sigma(y)^{(\tilde{R})}$ (see 3.1.2). Hence for all $y \in B$, if z is a real constructed by *Player II* in an arbitrary infinite run compatible with $\Sigma^*(y)$ (for example $a_n = \emptyset$ for all n), then it follows from the definition of B that the R-branch $\widehat{(y,z)}$ belongs to F hence the *Ext*-branch (y, z) belongs to X and $y = \pi(y, z) \in \pi(X)$.

To prove that $Y \subset B$ suppose by contradiction that $y \in Y$ and that for some position $u \in \Sigma^*_{II}(y)$ with $\chi(u) \in 2^m \times 2^m$ we have $N_{\chi(u)} \cap F = \emptyset$, and let $\bar{u}$ be the unique finite run in the S-game $\Sigma(y)$ defined by u (see 3.1.2). Pick now any $\alpha \in \omega^\omega$ such that $(y, \alpha) \in \lceil T \rceil$, and consider an infinite run in the game S-game $\Sigma(y)$, extending $\bar{u}$, and in which at his $(m+k+1)^{\text{th}}$ move *Player II* plays $(z_{|n_{m+k+1}}, \alpha_{|k+1})$, where n_{m+k+1} is the length of the position u_{m+k+1} played by *Player I* at his $(m+k+1)^{th}$ move. This defines an infinite run $\bar{v}$ in the initial game G: If y is the real constructed by *Player I* in this run, and if $\widehat{(y,z)}$ is the unique R-branch contained in (y, z) then $\widehat{(y,z)} \notin F$ hence $(y, z) \notin X$. So, since $a_{n_{p+1}}$ extends a_{n_p} for $p > m$, *Player I* wins the run $\bar{v}$ against τ, which is impossible. □

This finishes the proof of Theorem 3.3.1. □

REMARK 3.3.4. Unlike in Theorem 3.2.4, or even in the classical Suslin separation Theorem, where one separates two analytic sets by a Borel set, in Theorem 3.3.1 we separate an analytic set Y from a *coanalytic* set $Z := 2^\omega \setminus \pi(X)$ by a Borel set; but of course the pair (Y, Z) is not arbitrary.

In this respect one can say that the statement of Theorem 3.2.4 is more elementary than Theorem 3.3.1. Notice that this is also the case for the proofs given above of these results where despite some obvious similarity, the first uses simply a closed game, while the second is based on the determinacy of a Borel game. As a matter of fact H. Friedman showed ([**8**]) that the statement of Theorem 3.3.1 for an arbitrary Borel set interprets the existence of uncountably many cardinals; in particular one cannot hope to prove Theorem 3.3.1 by using only the determinacy of some closed game in the arguments.

CHAPTER 4

Borel Liftings of Borel Sets

In this chapter we shall prove two independent results about Borel liftings: the first one will enable us to solve Ostrovsky's Problem and the second one will answer a question of H. Friedman.

4.1. Borel liftings of bounded rank

For the solution of Ostrovsky's Problem we need to extend Theorem 3.3.1 in several directions: 1) by strengthening the conclusion; 2) by dealing with Borel liftings of fixed rank, instead of continuous liftings; 3) by working in arbitrary separable metrizable spaces. Each of these extensions is non-trivial and necessitates some additional new arguments. For practical reasons we shall work with the notion of *section* rather than the notion of *lifting*.

We recall that the descriptive complexity of a separable metrizable space is to be understood relatively to some (any) metric compactification of the space.

LEMMA 4.1.1. *For all $\xi < \omega_1$ there exists a Borel mapping $F : B \to 2^\omega$ of class ξ with $\mathbf{\Pi}^0_{1+\xi+1}$ domain $B \subset 2^\omega$ such that any Borel mapping $\varphi : A \to 2^\omega$ of class ξ with arbitrary domain $A \subset 2^\omega$ admits a factorization $\varphi = F \circ \psi$ where $\psi : A \to 2^\omega$ is Lipschitz and $\psi(A) \subset B$.*

PROOF. Fix any $\mathbf{\Sigma}^0_{1+\xi}$-complete set $U \subset 2^\omega$, set $U_0 = U \times 2^\omega$ and $U_1 = 2^\omega \times U$, and fix a pair (V_0, V_1) of disjoint $\mathbf{\Sigma}^0_{1+\xi}$ sets reducing the pair (U_0, U_1) of $\mathbf{\Sigma}^0_{1+\xi}$ sets in $2^\omega \times 2^\omega$ (i.e. $V_0 \subset U_0$, $V_1 \subset U_1$, $V_0 \cup V_1 = U_0 \cup U_1$ and $V_0 \cap V_1 = \emptyset$). Then $V = U_0 \cup U_1 = V_0 \cup V_1$ is a $\mathbf{\Sigma}^0_{1+\xi}$ subset of $2^\omega \times 2^\omega$ and $B = V^\omega$ is a $\mathbf{\Pi}^0_{1+\xi+1}$ subset of $(2^\omega \times 2^\omega)^\omega \approx 2^\omega$. So if f denotes the characteristic function of V_0 in V then clearly f is of class ξ hence the mapping $F : B \mapsto 2^\omega$ defined by $F((x_n)_{n\in\omega}) = (f(x_n))_{n\in\omega}$ is also Borel of class ξ.

Now consider any Borel mapping $\varphi : A \to 2^\omega$ of class ξ; then each coordinate mapping of φ defines a partition (A^0_n, A^1_n) of A into two relatively $\mathbf{\Delta}^0_{1+\xi}$ sets. So for $i = 0, 1$, we can write $A^i_n = A \cap U^i_n$ with $U^i_n \in \mathbf{\Sigma}^0_{1+\xi}$; and since $U_i \notin \mathbf{\Pi}^0_{1+\xi}$ Wadge's games $G_W(U^i_n, U_i)$ are won by Player II (see [**11**], 21.E). Thus there exist Lipschitz mappings $\psi^i_n : 2^\omega \to 2^\omega$ reducing U^i_n to U. Hence for all $x \in A^i_n$:

$$(\psi^0_n(x), \psi^1_n(x)) \in U_i \setminus U_{1-i} \subset V_i \subset V$$

and if we define $\psi : A \to B$ by:

$$\psi(x) = \Big((\psi^0_n(x), \psi^1_n(x))\Big)_{n\in\omega} \in V^\omega = B$$

then it is easy to check that ψ is Lipschitz and satisfies $\varphi = F \circ \psi$. □

LEMMA 4.1.2. *For all $\xi < \omega_1$ and all Borel sets $X \subset 2^\omega$ there exists a Borel mapping $G : X^* \to X$ of class ξ with Borel domain $X^* \subset 2^\omega$ such that any Borel* *mapping* $\varphi : K \to X$ *of class* ξ *with domain* $K \subset 2^\omega$, *admits a factorization* $\varphi = G \circ \varphi^*$ *where* $\varphi^* : K \to X^*$ *is Lipschitz.*

PROOF. Consider some Borel set $X \subset 2^\omega$. If $F : B \to 2^\omega$ is the Borel mapping of class ξ given by Lemma 4.1.1, set : $X^* := F^{-1}(X)$ and let $G := F_{|X^*}$.

If $\varphi : K \to X$ is Borel of class ξ with domain $K \subset 2^\omega$, then by Lemma 4.1.1 there exists a Lipschitz mapping $\varphi^* : K \to B \subset 2^\omega$ such that $\varphi = F \circ \varphi^*$. For all $\alpha \in K$, since $F(\varphi^*(\alpha)) = \varphi(\alpha) \in X$ then $\varphi^*(\alpha) \in X^*$ and φ^* can be viewed as a Lipschitz mapping from K to X^*. □

THEOREM 4.1.3. *Let* $\pi : 2^\omega \times 2^\omega \to 2^\omega$ *denote the canonical projection on the first factor,* $X \subset 2^\omega \times 2^\omega$ *be* $\mathbf{\Pi}^0_{1+\xi}$ $(\xi \geq 1)$, $Y \subset 2^\omega$ *be* $\mathbf{\Sigma}^1_1$, *and* η *be a countable ordinal. If any compact subset of* Y *admits a Borel lifting of class* η *in* X *then there exists a* $\mathbf{\Pi}^0_{1+\eta+\xi}$ *set* C *such that* $Y \subset C \subset \pi(X)$.

In particular if any compact subset of $\pi(X)$ *admits a Borel lifting of class* η *in* X *then* $\pi(X)$ *is Borel of class* $\mathbf{\Pi}^0_{1+\eta+\xi}$.

PROOF. Applying Lemma 4.1.1, we can find a Borel mapping $F : B \to 2^\omega$ of class η with $\mathbf{\Pi}^0_{1+\eta+1}$ domain $B \subset 2^\omega$ such that any Borel mapping $\varphi : A \to 2^\omega$ of class η admits a factorization $\varphi = F \circ \psi$ with $\psi : A \to B$ Lipschitz. Let

$$X^* := \{(\alpha, \beta) \in 2^\omega \times B : (\alpha, F(\beta)) \in X\}$$

Since the mapping : $(\alpha, \beta) \mapsto (\alpha, F(\beta))$ is of class η, X^* is $\mathbf{\Pi}^0_{1+\eta+\xi}$. Clearly $\pi(X^*) = \pi(X)$. Moreover, if K is a compact subset of Y, there is a Borel lifting $\varphi : K \to 2^\omega$ of class η in X, which by Lemma 4.1.1 admits a factorization $\varphi = F \circ \psi$ with $\psi : K \to B$ Lipschitz. Then for every $\alpha \in K$, $(\alpha, \psi(\alpha)) \in X^*$, hence ψ is a Lipschitz lifting of K in X^*. It follows then from Theorem 3.3.1 that there exists a $\mathbf{\Pi}^0_{1+\eta+\xi}$ set $C \subset 2^\omega$ such that $Y \subset C \subset \pi(X^*) = \pi(X)$. □

LEMMA 4.1.4. *Let* $X \subset 2^\omega \times 2^\omega$ *and* $Y \subset 2^\omega$ *be Borel sets. If any compact subset of* Y *admits a Lipschitz lifting in* X *then* Y *admits a Lipschitz lifting in* X.

PROOF. Consider the game G_0 in which:

– at his $(n+1)^{\text{th}}$ move Player I chooses $s_n \in 2^{(n+1)}$ such that: $s_m \prec s_n$ for all $m < n$

– at his $(n+1)^{\text{th}}$ move Player II chooses $t_n \in 2^{(n+1)}$ such that: $t_m \prec t_n$ for all $m < n$

Thus again in an infinite run in G_0, Player I constructs a real $y = \bigcup_n s_n \in 2^\omega$ and Player II constructs a real $z = \bigcup_n t_n \in 2^\omega$; and Player II wins the run if:

$$y \notin Y \quad \text{or} \quad (y, z) \in X$$

Since X *and* Y are Borel then G_0 is a Borel game hence determined. Obviously any winning strategy for Player II in G_0 provides a Lipschitz $\varphi : 2^\omega \to 2^\omega$ whose restriction to Y defines a lifting of Y in X. The lemma follows then from the next claim.

CLAIM . *Player I has no winning strategy in* G_0.

<u>*Proof*</u>: Suppose that σ is a winning strategy for Player I in G_0, and consider the set K of all reals constructed by Player I in all infinite runs compatible with σ. Since at each move Player II has only finitely many (in fact two) possible choices, it is clear that K is compact; and it follows from the win condition that $K \subset Y$.

Fix a Lipschitz mapping $\varphi : K \to 2^\omega$ such that $\{(y, \varphi(y))\,;\ y \in K\} \subset X$, and let $\phi : T_K \to \mathrm{Seq}(A)$ be the Lipschitz extension mapping of φ. Notice that if s is a move made by Player I in a run in G_0 compatible with σ then it follows from the definition of K that $N_s \cap K \neq \emptyset$ hence $s \in T_X$, and $\phi(s)$ is well defined. Consider then the infinite run in G_0 in which Player I follows σ and Player II answers $\phi(s)$ to any move s; and let y_0 and z_0 be the reals constructed by Player I and Player II in this run. Then $z_0 = \varphi(y_0)$, and by definition of K we have $y_0 \in K$ hence $(y_0, z_0) \in X$, and Player II wins the run, which gives the contradiction. ◇

□

THEOREM 4.1.5. *Let $\xi \geq 0$ be a countable ordinal, and $f : X \to Y$ be a Borel onto mapping between two zero-dimensional separable metrizable spaces with Borel domain X. If f admits a Borel section of class ξ on any compact subset of Y, then Y is Borel and f admits a Borel section of class ξ on Y.*

PROOF. Since X and Y are zero-dimensional we can and do assume they are subsets of 2^ω. Let $G : X^* \to X$ be the Borel mapping of class ξ given by Lemma 4.1.2 and consider the Borel mapping $f^* = f \circ G : X^* \to Y$.

By hypothesis, on any compact subset K of Y, f admits a Borel section $\varphi : K \to Y$ of class ξ, and applying Lemma 4.1.1 we can find a Lipschitz mapping $\varphi^* : K \to X^*$ such that $\varphi = G \circ \varphi^*$; then $f^* \circ \varphi^* = f \circ G \circ \varphi^* = f \circ \varphi = Id_K$ and φ^* is a Lipschitz section for f^*.

Now notice that the set $X' = \{(y, z) \in 2^\omega \times X^* : y = f^*(z)\} \subset 2^\omega \times 2^\omega$ is Borel and that $Y = \pi(X')$; moreover by the observations above any compact subset of Y admits a Lipschitz lifting in X'. It follows then from Theorem 3.3.1 that Y is also Borel, and applying now Lemma 4.1.4 we get a Lipschitz lifting of Y in X' which defines a Lipschitz section $\psi^* : Y \to X^*$ for f^*. It follows then that $\psi = G \circ \psi^*$ is a Borel section of class ξ for the function f. □

REMARKS 4.1.6. a) For non-zero-dimensional spaces, Theorem 4.1.5 is false for $\xi = 0$. For example if $X = \{(x, n) \in \mathbb{R} \times \omega : x \neq n\}$ and $Y = \mathbb{R}$ then:

i) Any compact subset of Y admits clearly a constant lifting in X.

ii) But Y has no continuous lifting in X: indeed if φ were such a lifting of Y in X, then by connectedness $\varphi(Y)$ would be reduced to a singleton $\{p\}$ and φ would be constant; and this is impossible since $(p, \varphi(p)) = (p, p) \notin X$.

b) Let $X \subset 2^\omega \times 2^\omega$ be a $\mathbf{\Pi}^0_{1+\xi}$ set and suppose that $Y \subset 2^\omega$ admits in X a Borel lifting $\varphi : Y \to 2^\omega$ of Baire class η. By classical results φ admits then an extension $\tilde{\varphi} : \tilde{Y} \to 2^\omega$ with $\mathbf{\Pi}^0_{1+\eta+1}$ domain $\tilde{Y}$ and which is also of Baire class η. Hence the set $C := \tilde{\varphi}^{-1}(X)$ is a $\mathbf{\Pi}^0_{1+\xi+\eta}$ set which separates Y from $\pi(X)$ and on which $\tilde{\varphi}$ is still a Borel lifting in X of Baire class η. It follows from these observations that Theorem 4.1.3 and Theorem 4.1.5 above are immediate consequences of the following more general statement:

"*Let X be a Borel subset of $2^\omega \times 2^\omega$ and $Y \subset 2^\omega$ be $\mathbf{\Sigma}^1_1$. If any compact subset of Y admits a Borel lifting in X of class η then Y admits a Borel lifting in X of class η*"

This statement is easily provable under $Det(\mathbf{\Sigma}^1_1)$. However we were able to prove that its validity for $\eta = 0$ and X $\mathbf{\Pi}^0_2$ is already equivalent to $Det(\mathbf{\Sigma}^1_1)$; the arguments are too far away from the spirit of the present work and we do not develop them here.

Extension to metric spaces:

We now extend the previous theorem to the frame of arbitrary separable metrizable spaces. In next two lemmas L, M, N denote arbitrary separable metrizable spaces. The first lemma is just Lemma 3 of [**22**] :

LEMMA 4.1.7. *If $q : M \to N$ is perfect, then for any Borel mapping $\varphi : M \to L$ of Baire class $\xi \geq 1$ there exists a Borel section σ of q of Baire class 1 such that the mapping $\psi := \varphi \circ \sigma$ is of Baire class ξ.*

In particular in Lemma 4.1.7, the mapping φ admits a factorization $\varphi = \psi \circ q$ with $\psi : N \to L$ of Baire class ξ. We now prove a dual factorization result. For this we need to introduce a new notion.

We shall say that a mapping $p : M \to N$ is completely open if:

(i) p is onto, continuous and open.

(ii) There exists some fixed compatible metric on M for which each fiber $p^{-1}(y)$ is complete.

Notice that if M is Polish then by the continuity of p, condition (ii) follows from (i).

The following lemma is probably well known.

LEMMA 4.1.8. *For any Polish space P there exist*

a) a closed subset P_0 of ω^ω and an open mapping p from P_0 onto P,

b) a closed subset P_1 of ω^ω and a perfect mapping q from P_1 onto P.

PROOF. *a)* Fix a complete metric on P and define inductively open sets $(U_s)_{s \in \mathrm{Seq}(\omega)}$ of P such that $U_\emptyset = P$ and satisfying for every $s \in \mathrm{Seq}(\omega)$:

(i) diam$(U_{s^\frown n}) \leq 2^{-|s|}$ and $\overline{U_{s^\frown n}} \subset U_s$ for every $n \in \omega$.

(ii) the family $(U_{s^\frown n})_n$ is an open cover of U_s.

Then $P_0 := \{\alpha \in \omega^\omega : \forall s \prec \alpha \quad U_s \neq \emptyset\}$ is a closed subset of ω^ω. Moreover for each $\alpha \in P_0$ there is a unique point $p(\alpha)$ in P such that $\{p(\alpha)\} = \bigcap_{s \prec \alpha} \overline{U_s} = \bigcap_{s \prec \alpha} U_s$. The function $p : P_0 \to P$ is continuous and satisfies $p(P_0 \cap N_s) = U_s$ for each s, hence is open and onto.

b) Define in a similar way closed sets $(H_s)_{s \in \mathrm{Seq}(\omega)}$ of P such that $H_\emptyset = P$ and satisfying for every $s \in \mathrm{Seq}(\omega)$:

(i) diam$(H_{s^\frown n}) \leq 2^{-|s|}$ and $H_{s^\frown n} \subset H_s$ for every $n \in \omega$.

(ii) the family $(H_{s^\frown n})_n$ is an locally finite cover of H_s.

Indeed if H_s is defined, the open cover $\left(H_s \cap B(x, 2^{-|s|-1})\right)_{x \in H_s}$ of H_s admits a locally finite refinement $(V_n)_{n \in \omega}$ and we can define $H_{s^\frown n} := \overline{V_n}$. Then $P_1 := \{\alpha \in \omega^\omega : \forall s \prec \alpha \quad H_s \neq \emptyset\}$ is a closed subset of ω^ω. As above for each $\alpha \in P_1$ there is a unique point $q(\alpha)$ in P such that $\{q(\alpha)\} = \bigcap_{s \prec \alpha} H_s$. The function $q : P_1 \to P$ is continuous and satisfies $q(P_1 \cap N_s) = H_s$ for each s, hence is onto. Finally, for every compact subset K of P, it follows from condition *ii)* that the set $\{s \in \mathrm{Seq}(\omega) : |s| = k \text{ and } K \cap H_s \neq \emptyset\}$ is finite for each integer k, hence that $p^{-1}(K)$ is a compact subset of ω^ω contained in P_1 ; and this shows that q is perfect. □

LEMMA 4.1.9. *If $p : M \to N$ is completely open, then any Borel mapping $\varphi : L \to N$ of Baire class $\xi \geq 1$, admits a factorization $\varphi = p \circ \psi$ with $\psi : L \to M$ of Baire class ξ.*

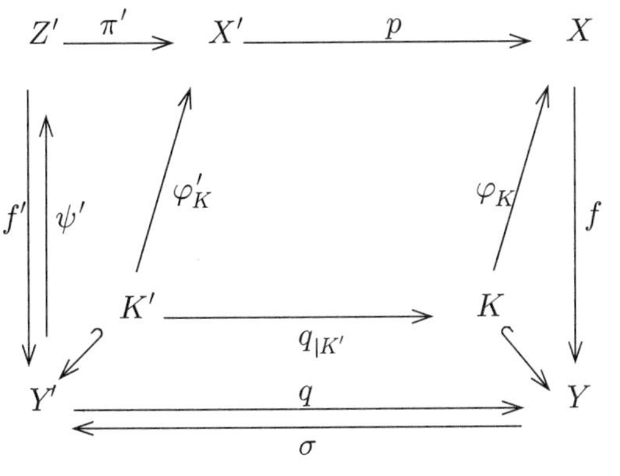

FIGURE 4.1

PROOF. Fix a metric on M for which all p-fibers are complete. One can easily construct by induction a family $(V_s)_{s\in\mathrm{Seq}(\omega)}$ of open sets in M such that $V_\emptyset = M$, satisfying for all $s \in \mathrm{Seq}(\omega)$:

(i) $(V_{s^\frown k})_{k\in\omega}$ is a cover of V_s by open sets of diameter $\leq 2^{-|s|}$

(ii) For any $s \in \omega^n$ and any $k \in \omega$, $\overline{V_{s^\frown k}} \subset V_s$.

Then for all s , $W_s := p(V_s)$ is open in N, hence for all $n \geq 0$, $(\varphi^{-1}(W_s))_{s\in\omega^n}$ is a cover of L by $\mathbf{\Sigma}^0_{1+\xi}$ sets, and by $\mathbf{\Sigma}^0_{1+\xi}$ reduction we can find a partition $(A_s)_{s\in\omega^n}$ of L by $\mathbf{\Delta}^0_{1+\xi}$ sets, which is finer than the cover $(\varphi^{-1}(W_s))_{s\in\omega^n}$ and satisfying $A_s = \bigcup_{k\in\omega} A_{s^\frown k}$ for all $s \in \mathrm{Seq}(\omega)$. Hence for all $x \in L$, there exists a unique $\alpha \in \omega^\omega$ such that $x \in \bigcap_{s\prec\alpha} A_s$, and it follows from the completeness assumption on the p-fibers and the diameter condition on the V_s that $p^{-1}(\varphi(x)) \cap \bigcap_{s\prec\alpha} V_s$ is a singleton $\{\psi(x)\}$. This defines a mapping $\psi : L \to M$ such that $p(\psi(x)) = \varphi(x)$ and $\psi^{-1}(V) = \bigcup\{\mathbb{A}_s : V_s \subset V\}$ is $\mathbf{\Sigma}^0_{1+\xi}$ for all open subset V of M; so ψ is Borel of Baire class ξ and $\varphi = p\circ\psi$. □

THEOREM 4.1.10. *Let $\xi > 0$ be a countable ordinal, and $f : X \to Y$ be a Borel mapping between two separable metrizable spaces with Borel domain X. If the restriction of f to any compact subset of Y admits a Borel section of class ξ in X, then Y is Borel and f admits a Borel section of class ξ on Y.*

PROOF. Embed X and Y into two Polish spaces P and Q, and fix two continuous and onto mappings $p_0 : P_0 \to P$ and $q_0 : Q_0 \to Q$, with P_0 and Q_0 closed subsets of ω^ω, p_0 open and q_0 perfect, as in Lemma 4.1.8. Set: $X' = p_0^{-1}(X) \subset P_0 \subset \omega^\omega$, $Y' = q_0^{-1}(Y) \subset Q_0 \subset \omega^\omega$ and let $p : X' \to X$ and $q : Y' \to Y$ denote the restrictions of p_0 and q_0 to X' and Y' respectively. It is clear then that p is completely open and q is perfect. Finally consider the Borel set:

$$Z' = \{(\alpha, \beta) \in P_0 \times Q_0 \ : \ p_0(\alpha) \in X \text{ and } q_0(\beta) = f(p_0(\alpha))\}$$

Notice that the projections of Z' on the first and second factor are contained in X' and Y' respectively. Finally let $\pi' : Z' \to X'$ and $f' : Z' \to Y'$ denote the restrictions to Z' of the projections on the first and the second factor.

CLAIM . *On any compact subset of Y', f' admits a Borel section of class ξ.*

Proof: Let K' be a compact subset of Y'; then applying the hypothesis of the theorem to the compact set $K = q(K') \subset Y$ we get a Borel section $\varphi_K : K \to X$ of class ξ for f. Now since p is completely open then by Lemma 4.1.9, we can find a Borel mapping $\varphi'_K : K' \to X'$ of class ξ such that $\varphi_K \circ q_{|K'} = p \circ \varphi'_K$, where $q_{|K'}$ denotes the restriction of q to K'. Finally for $\beta \in K'$ define $\psi'_K(\beta) = (\alpha, \beta)$ for $\alpha := \varphi'_K(\beta)$. Since $\alpha \in X'$ then $p_0(\alpha) = p(\alpha) \in X$; moreover since φ_K is a section for f on $q(K')$ then $f(p_0(\alpha)) = f(p(\varphi'_K(\beta))) = f(\varphi_K(q(\beta))) = q(\beta) = q_0(\beta)$; hence $(\alpha, \beta) \in Z'$ and so ψ'_K is a a Borel section of class ξ for f' on K'. ◇

Since Z' and Y' are zero-dimensional then by Theorem 4.1.5 there exists a Borel section $\psi' : Y' \to Z'$ of class ξ for f'. Set $\varphi' := \pi' \circ \psi'$; then $p \circ \varphi' : Y' \to X$ is of class ξ, and since q is perfect then by Lemma 4.1.7, we can find a Borel section σ of q such that $p \circ \varphi' \circ \sigma$ is a Borel mapping $\varphi : Y \to X$ of class ξ. Then for all $y \in Y$, $\beta := \sigma(y)$ satisfies $f' \circ \psi'(\beta) = \beta$, hence $f \circ \varphi(y) = q(\beta) = y$; this shows that φ is a section for f. □

4.2. Borel liftings of unbounded rank

In this section we shall consider the following statement:

Dom: "$\forall\, \alpha \in \omega^\omega$, $\omega^\omega \cap L(\alpha)$ *is* $\leq^*$*-bounded in* ω^ω"

where $\leq^*$ denotes eventual domination on ω^ω. We shall also consider later on its lightface version:

Dom: "$\omega^\omega \cap L$ *is* $\leq^*$*-bounded in* ω^ω"

If one replaces in the statement of Theorem 4.1.10 "Borel of Baire class ξ" by "Borel" (of unbounded rank), then the result is no more true, even if one assumes both X and Y Borel. In fact in ([**8**], Section 6) H. Friedman constructs a Borel subset X of $\omega^\omega \times \omega^\omega$ of total projection ($\pi(X) = \omega^\omega$) such that any compact subset of ω^ω admits a Borel lifting but for which the existence of Borel liftings with domain ω^ω implies that **Dom** holds. Friedman also asks ([**8**], remark after Proposition 7) whether the following converse holds:

Problem: *Assume that* **Dom** *holds.*
Let $X \subset \omega^\omega \times \omega^\omega$ and $Y \subset \omega^\omega$ be Borel, and suppose that any compact subset of Y admits a Borel lifting in X. Does Y admit a Borel lifting in X ?

In fact as observed by Friedman in [**8**], it follows from the results of [**3**] that the answer is positive if moreover Y is supposed to be $\mathbf{\Pi}^0_2$. We now settle the general case that we again state in the setting of arbitrary separable metrizable spaces.

THEOREM 4.2.1. *Assume* **Dom**. *Let $f : X \to Y$ be a Borel mapping between two Borel metrizable separable spaces. If the restriction of f to any compact subset of Y admits a Borel section, then f admits a Borel section on Y.*

PROOF. For the proof we will need the following analog of Lemma 4.1.1 in which we consider Borel mappings of arbitrary class:

LEMMA 4.2.2. *There exists a $\mathbf{\Pi}^1_1$-recursive mapping $F : P \to \omega^\omega$ with $\mathbf{\Pi}^1_1$ domain $P \subset 2^\omega$ such that any Borel mapping $\varphi : A \to \omega^\omega$, with arbitrary domain $A \subset 2^\omega$, admits a factorization $\varphi = F \circ \psi$ where $\psi : A \to 2^\omega$ is continuous and $\psi(A) \subset P$.*

PROOF. Fix any $\mathbf{\Pi}^1_1$-complete set $U \subset 2^\omega$, set $U_0 = U \times 2^\omega$ and $U_1 = 2^\omega \times U$, and fix a pair (V_0, V_1) of disjoint $\mathbf{\Pi}^1_1$ sets reducing the pair (U_0, U_1) of $\mathbf{\Pi}^1_1$ sets in $2^\omega \times 2^\omega$. Then $V = U_0 \cup U_1 = V_0 \cup V_1$ is a $\mathbf{\Pi}^1_1$ subset of $2^\omega \times 2^\omega$ and $P_1 := V^\omega$ is a $\mathbf{\Pi}^1_1$ subset of $(2^\omega \times 2^\omega)^\omega \approx 2^\omega$. So if f denotes the characteristic function of V_0 in V then clearly f is $\mathbf{\Pi}^1_1$-recursive with $\mathbf{\Pi}^1_1$ domain hence the mapping $F_1 : P_1 \to 2^\omega$ defined by $F_1((x_n)_{n\in\omega}) = (f(x_n)_{n\in\omega})$ is also $\mathbf{\Pi}^1_1$-recursive. Let j be a homeomorphic embedding of ω^ω into 2^ω and set $P := \{\alpha \in P_1 : F_1(\alpha) \in j(\omega^\omega)\}$ then $F := j^{-1}\circ F_1 : P \to \omega^\omega$.

Now consider any Borel mapping $\varphi : A \to \omega^\omega$. Then each coordinate mapping of $j\circ\varphi$ defines a partition (A^0_n, A^1_n) of A into two relatively $\mathbf{\Delta}^1_1$ sets. So for $i = 0, 1$, we can write $A^i_n = A \cap U^i_n$ with $U^i_n \in \mathbf{\Pi}^1_1$; and since U_i is $\mathbf{\Pi}^1_1$-complete, there exist continuous mappings $\psi^i_n : 2^\omega \to 2^\omega$ reducing U^i_n to U. Hence for all $x \in A^i_n$:

$$(\psi^0_n(x), \psi^1_n(x)) \in U_i \setminus U_{1-i} \subset V_i \subset V$$

and if we define $\psi : A \to P_1$ by:

$$\psi(x) = ((\psi^0_n(x), \psi^1_n(x))_{n\in\omega} \in V^\omega = P_1$$

then one easily checks that ψ is continuous, $\psi(A) \subset P$ and $j\circ\varphi = F_1\circ\psi$, hence $\varphi = j^{-1}\circ F_1\circ\psi = F\circ\psi$. □

<u>*Proof of Theorem 4.2.1*</u> : We equip as usual the hyperspace $\mathscr{K}(2^\omega \times 2^\omega)$ of all compact subsets of $2^\omega \times 2^\omega$ with the Vietoris topology. Let

$$\mathscr{G} := \{H \in \mathscr{K}(2^\omega \times 2^\omega) : \forall \alpha, \beta, \beta' \in 2^\omega, \ ((\alpha,\beta) \in H \text{ and } (\alpha,\beta') \in H\,) \to \beta = \beta'\,\}$$

thus $\mathscr{G}$ is the set of all compact subsets of $2^\omega \times 2^\omega$ which are the graphs of partial mappings (necessarily continuous with compact domain). For every integer n if we set

$$A_n := \{(K, \alpha, \beta, \beta') \in \mathscr{K}(2^\omega \times 2^\omega) \times (2^\omega)^3 : (\alpha, \beta) \in K \text{ and } (\alpha, \beta') \in K \text{ and } d(\beta, \beta') \geq 2^{-n}\}$$

then clearly each A_n is compact and

$$\mathscr{G} = \mathscr{K}(2^\omega \times 2^\omega) \setminus \bigcup_n \{K : \exists(\alpha, \beta, \beta') \in (2^\omega)^3 : (K, \alpha, \beta, \beta') \in A_n\}$$

which shows $\mathscr{G}$ is $\mathbf{\Pi}^0_2$. Consider now the set

$$\mathscr{F} := \{(\alpha, H) \in 2^\omega \times \mathscr{G} : \ \exists\beta \in 2^\omega \quad (\alpha, \beta) \in H\,\}$$

which is also $\mathbf{\Pi}^0_2$, since a relatively closed subset of $2^\omega \times \mathscr{G}$. For any $(\alpha, H) \in \mathscr{F}$ let $\rho(\alpha, H)$ denote the unique $\beta \in 2^\omega$ such that $(\alpha, \beta) \in H$. This defines a mapping $\rho : \mathscr{F} \to 2^\omega$ whose graph is clearly a closed subset of $\mathscr{F} \times 2^\omega$, hence ρ is continuous.

We can assume that Y is uncountable, for otherwise any section of f would be Borel; and since f is onto then X is uncountable too. So we can fix two continuous bijections $\varphi : \omega^\omega \to X$ and $\psi : \omega^\omega \to Y$, and it is a well known fact that φ^{-1} and ψ^{-1} are then Borel.

Let P and F be the $\mathbf{\Pi}^1_1$ set and the $\mathbf{\Pi}^1_1$-recursive function given by Lemma 4.2.2. Embed ω^ω as a $\mathbf{\Pi}^0_2$ subset of 2^ω and consider the $\mathbf{\Pi}^1_1$ set

$$Z := \{(\alpha, H) \in \mathscr{F} : \rho(\alpha, H) \in P \text{ and } \psi(\alpha) = f\circ\varphi\circ F\circ\rho(\alpha, H))\}$$

CLAIM . *Any compact subset of ω^ω admits a constant lifting in Z.*

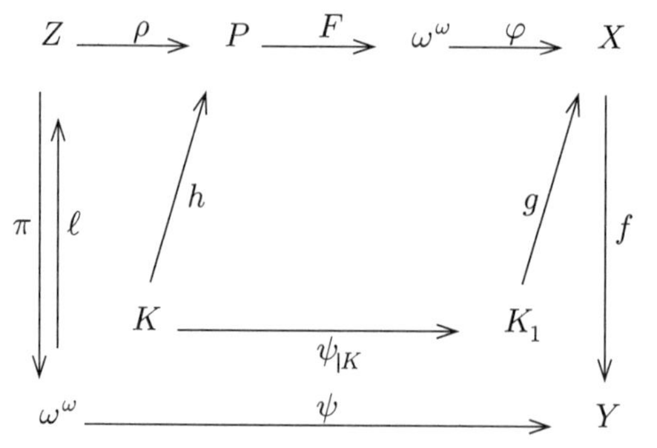

FIGURE 4.2

If K is a compact subset of ω^ω, then $K_1 := \psi(K)$ is a compact subset of Y. By hypothesis there is a Borel section g of f on K_1. So $\varphi^{-1}\circ g\circ\psi_{|K} : K \to \omega^\omega$ is Borel so admits by Lemma 4.2.2 a factorization $\varphi^{-1}\circ g\circ\psi = F\circ h$ with $h : K \to P$ continuous.

If H is the graph of h then $H \in \mathscr{G}$, and for all $\alpha \in K$, $(\alpha, H) \in \mathscr{F}$; moreover if $\beta := h(\alpha) = \rho(\alpha, H)$ then $\varphi^{-1}\circ g\circ\psi(\alpha) = F\circ h(\alpha) = F(\beta)$ hence $g\circ\psi(\alpha) = \varphi\circ F(\beta)$, and since g is a section for f on K_1 then $\psi(\alpha) = f\circ g\circ\psi(\alpha) = f\circ\varphi\circ F(\beta) = F\circ\rho(\alpha, H)$; and this shows that $(\alpha, H) \in Z$. Hence the constant mapping $\alpha \mapsto H$ is a lifting of K in Z. $\diamond$

Since **Dom** holds, it follows then from ([**3**], Theorems 5.2 and 6.3) that ω^ω admits a continuous lifting in Z, so there exists a continuous mapping $\ell : \alpha \mapsto (\alpha, H)$ from ω^ω to Z which is a section for the canonical projection π from Z onto ω^ω. Then it follows from the very definition of Z that the mapping $\sigma := \varphi\circ F\circ\rho\circ\ell\circ\psi^{-1} : Y \to X$ is a section of f which is $\mathbf{\Pi}^1_1$-recursive by Lemma 4.2.2; and since Y is $\mathbf{\Sigma}^1_1$ then σ is Borel. $\square$

4.3. Solution to Ostrovsky's problem

THEOREM 4.3.1. *Let $f : X \to Y$ be a compact covering mapping between two separable metrizable spaces.*

If X is Borel then Y is Borel and of the same additive or multiplicative Baire class.

PROOF. Let $\hat{X}$ and $\hat{Y}$ be metrizable compactifications of X and Y, $p_0 : 2^\omega \to \hat{X}$ and $q_0 : 2^\omega \to \hat{Y}$ be continuous onto mappings. Let $X' = p_0^{-1}(X)$ and $Y^* = q_0^{-1}(Y)$. Then the restrictions p of p_0 to X' and q of q_0 to Y^* are onto and perfect. Let

$$X^* := \{(\alpha, \beta) \in 2^\omega \times X' : q_0(\alpha) = f\circ p(\beta)\}$$

If X is $\mathbf{\Pi}^0_{1+\xi}$, then X^* is also $\mathbf{\Pi}^0_{1+\xi}$. Let K^* be a compact subset of Y^*, $K := q(K^*)$ is a compact subset of Y, and there is a compact subset L of X such that $f(L) = K$. Then $L' := p^{-1}(L)$ is compact in X', and $L^* = \{(\alpha, \beta) \in K^* \times L' : q_0(\alpha) = f\circ p(\beta)\}$ is a compact subset of X^* such that $\pi_{|X^*}(L^*) = K^*$.

Then the restriction of $\pi_{|X^*}$ to L^* and K^* is perfect, hence by classical results K^* admits a Borel lifting in X^* of Baire class one. It follows then from Theorem 4.1.3 that Y^* is $\mathbf{\Pi}^0_{2+\xi}$. And since q is perfect, it follows from Lemma 4.1.7 that Y is $\mathbf{\Pi}^0_{2+\xi}$ too. This shows that Y is Borel and even gives a (non optimal) bound for the class of Y; but applying Theorem 8.6 of [**2**] one can conclude that Y is of the same additive or multiplicative class as X. □

REMARKS 4.3.2. a) Ostrovsky's Problem was initially considered for Baire classes. But it is clear from the arguments of Theorem 4.3.1 that the same result also holds for any Lavrentieff class $D_\eta(\mathbf{\Sigma}^0_\xi)$ or $\check{D}_\eta(\mathbf{\Sigma}^0_\xi)$ with $\xi \geq 2$.

b) Suppose that one works in the context of recursively presented Polish spaces, and that the mapping f (hence its domain X) is Δ^1_1. It is not clear in this case whether the set Y is also Δ^1_1. However inspecting the proof of Theorem 4.3.1, in particular noting that the winning strategy for *Player II* given by the proof of Theorem 3.3.1 has a code in L, since the game G_0 is Δ^1_1, one easily checks that in this case the set Y admits a Borel code in L.

4.4. Borel liftings in coanalytic sets

Let us recall the following notations from the Introduction, where Λ and Γ are arbitrary classes:

Lift(Λ,Γ) : *"For any $X \subset 2^\omega \times 2^\omega$ in Λ and any $Y \subset 2^\omega$ in Γ, if any compact subset of Y has a continuous lifting in X then Y has a continuous lifting in X"*

Lift$^{(\xi)}(\Lambda,\Gamma)$: *"For any $X \subset 2^\omega \times 2^\omega$ in Λ and any $Y \subset 2^\omega$ in Γ, if any compact subset of Y has a Borel lifting of class ξ in X then Y has a Borel lifting of class ξ in X"*

Lift$^{*}(\Lambda,\Gamma)$: *"For any $X \subset 2^\omega \times 2^\omega$ in Λ and any $Y \subset 2^\omega$ in Γ, if any compact subset of Y has a Borel lifting in X then Y has a Borel lifting in X"*

So *Lift*$^{(0)}(\Lambda,\Gamma)$ is just *Lift* (Λ,Γ), but since this statement is central in our study we shall use the simpler notation *Lift* (Λ,Γ).

Notice that by Theorem 4.1.10 *Lift*$^{(\xi)}(\mathbf{\Delta}^1_1,\mathbf{\Delta}^1_1)$ holds (in ZFC) for all ξ. But as we mentioned above, in [**8**] Friedman proves (Theorem 30) that *Lift*$^{(\star)}(\mathbf{\Delta}^1_1,\mathbf{\Pi}^0_2)$ implies **Dom**, hence combining this with Theorem 4.2.1 one gets:

$$\mathit{Lift}^{(\star)}(\mathbf{\Delta}^1_1,\mathbf{\Delta}^1_1) \iff \mathbf{Dom}$$

However when one considers liftings in $\mathbf{\Pi}^1_1$ sets the comparison of *Lift*$^{(\xi)}$ and *Lift*$^{(\star)}$ is inverted:

PROPOSITION 4.4.1. *For any class $\Gamma \subset \mathbf{\Pi}^1_1$, and any countable ordinal ξ:*

$$\mathit{Lift}(\mathbf{\Pi}^1_1,\Gamma) \implies \mathit{Lift}^{(\xi)}(\mathbf{\Pi}^1_1,\Gamma) \implies \mathit{Lift}^{(\star)}(\mathbf{\Pi}^1_1,\Gamma)$$

PROOF. Assume *Lift*$(\mathbf{\Pi}^1_1,\Gamma)$; and towards proving *Lift*$^{(\xi)}(\mathbf{\Pi}^1_1,\Gamma)$ fix $X \subset 2^\omega \times 2^\omega$ in $\mathbf{\Pi}^1_1$ and $Y \subset 2^\omega$ in Γ such that any compact subset of Y has a Borel lifting of class ξ in X.

Consider again the Borel mapping $F : B \to 2^\omega$ of class ξ, given by Lemma 4.1.1. Let $\mathscr{C}$ denote the set of all continuous mappings $\psi : K \to 2^\omega$ with compact domain $K \subset 2^\omega$. We identify $\mathscr{C}$ with a subset $\mathscr{G}$ of $\mathscr{K}(2^\omega \times 2^\omega)$ (by identifying each mapping $\psi \in \mathscr{G}$ with its graph $Gr(\psi)$). Since $\mathscr{G}$ is a $\mathbf{\Pi}^0_2$ subset of $\mathscr{K}(2^\omega \times 2^\omega)$ (see the proof of Theorem 4.2.1) $\mathscr{C}$ is a zero-dimensional Polish space, so we can identify $\mathscr{C}$ with a $\mathbf{\Pi}^0_2$ subset of 2^ω. Let $\mathscr{Z} := \{\psi \in \mathscr{C} : Gr(F \circ \psi) \subset X\} = \{\psi \in \mathscr{C} : Gr(\psi)) \subset G^{-1}(X)\}$ where $G : 2^\omega \times B \to 2^\omega \times 2^\omega$ denotes the Borel mapping $Id \times F$; then clearly $\mathscr{Z}$ is $\mathbf{\Pi}^1_1$.

On the other hand the set $\mathscr{X}_0 := \{(y, \psi) \in 2^\omega \times \mathscr{C} : y \in \mathrm{dom}(\psi)\}$ can be identified to a $\mathbf{\Pi}^0_2$ subset of the product space $2^\omega \times 2^\omega$, hence $\mathscr{X} := \mathscr{X}_0 \cap (2^\omega \times \mathscr{Z})$ is $\mathbf{\Pi}^1_1$; and it follows from the hypothesis that:

$$\forall K \text{ compact} \subset Y\,,\ \exists \psi \in \mathscr{Z} \text{ such that } K \times \{\psi\} \subset \mathscr{X}$$

so any compact subset of Y has a continuous (in fact constant) lifting in $\mathscr{X}$. Hence by $\mathit{Lift}(\mathbf{\Pi}^1_1, \Gamma)$ there exists a continuous mapping $\Psi^* : Y \to \mathscr{C}$ such that for all $y \in Y$, $\Psi^*(y)$ is a continuous mapping with compact domain, and range contained in B and such that $y \in \mathrm{dom}(\Psi^*(y))$; hence we can define a mapping $\psi^* : Y \to 2^\omega$ by

$$\psi^*(y) = F((\psi^*(y))(y))$$

which is clearly a a Borel lifting of class ξ in X.

This proves the first implication. The proof of the second implication is totally similar: one has only to use the mapping F given by Lemma 4.2.2 instead of the mapping given by Lemma 4.1.1. $\square$

REMARKS 4.4.2. a) Consider the following weak form of $\mathit{Lift}(\Lambda, \Gamma)$:

$\mathit{Lift}^{(00)}(\Lambda, \Gamma)$: *"For any $X_0 \subset 2^\omega \times 2^\omega$ in Δ^1_1, any $Y \subset 2^\omega$ in Γ, any $Z \subset 2^\omega$ in Λ, if for any compact subset K of Y there exists $z \in Z$ such that $K \times \{z\} \subset X_0$, then Y has a continuous lifting in $X := X_0 \cap (2^\omega \times Z)$."*

A simple inspection of the proof of Proposition 4.4.1 shows that what we really proved is the following stronger implication

$$\mathit{Lift}^{(00)}(\mathbf{\Pi}^1_1, \Gamma) \implies \mathit{Lift}^{(\xi)}(\mathbf{\Pi}^1_1, \Gamma)$$

which is significant even for $\xi = 0$.

b) One can also easily check that all arguments above (including the proofs of Lemma 4.1.1 and Lemma 4.2.2 as well as the previous remark) are effective. In particular we have the following "lightface" version of Proposition 4.4.1:

$$\mathit{Lift}^{(00)}(\Pi^1_1, \Gamma) \implies \mathit{Lift}^{(\xi)}(\Pi^1_1, \Gamma) \implies \mathit{Lift}^{(*)}(\Pi^1_1, \Gamma)$$

for any lightface class $\Gamma \subset \Pi^1_1$ and any recursive ordinal ξ.

4.4.3. Main result: From now on we focus on *Lift*. Before we announce the main result of the present work we review briefly previous known ones.

Let us first point out that if $\mathscr{P}$ denotes the class of all sets of reals then $\mathit{Lift}\,(\mathscr{P}, \mathbf{\Sigma}^0_2)$ holds in ZFC (we recall that the descriptive complexity is computed in some compactification, so here $\mathbf{\Sigma}^0_2$ means *σ-compact*). This unpublished result is a variation of Ostrovsky's result ([**20**]). We do not give the proof since the arguments are not in the spirit of this work. This is the only ZFC result about lifting properties in non-Borel sets. We then have the following equivalences:

$$(a) \qquad \mathit{Lift}(\mathbf{\Pi}^1_1, \mathbf{\Pi}^1_1) \iff \mathit{Det}(\mathbf{\Sigma}^1_1)$$

$$(b) \qquad \mathit{Lift}(\mathbf{\Pi}^1_1, \mathbf{\Pi}^0_2) \iff \mathbf{Dom}$$

$$(c) \qquad \mathit{Lift}(\mathbf{\Pi}^1_1, \mathbf{\Delta}^1_1) \iff \forall\alpha \in \omega^\omega,\ \aleph_1^{L(\alpha)} < \aleph_1$$

These equivalences are merely translations in the *Lift* language of the main results of [**2**], [**3**], [**4**], [**6**]. Moreover the arguments of the proofs of *(a)* and *(b)* are uniform and in fact we also have:

$$(a') \qquad \mathit{Lift}(\Pi^1_1, \Pi^1_1) \iff \mathit{Det}(\Sigma^1_1)$$

$$(b') \qquad \mathit{Lift}(\Pi^1_1, \Pi^0_2) \iff \mathit{Dom}$$

But the proof of *(c)*, which is much more elaborated than the proofs of *(a)* and *(b)*, does not give any lightface equivalence; though from [**4**] and the arguments of [**6**] one can get the following weak estimate:

$$\aleph^L_{\omega_1^{CK}} < \aleph_1 \implies \mathit{Lift}\,(\Pi^1_1, \Delta^1_1) \implies \aleph^L_1 < \aleph_1$$

and the main motivation of the present work was to give a lightface verion for *(c)*. We shall solve this problem by proving the following level by level equivalence:

$$(c'_\xi) \qquad \mathit{Lift}(\Pi^1_1, \Sigma^0_{1+\xi+1}) \iff \aleph^L_\xi < \aleph_1$$

where ξ is a recusive ordinal.

Both implications in (c'_ξ) are non trivial; but getting $\mathit{Lift}\,(\Pi^1_1, \Sigma^0_{1+\xi+1})$ from $\aleph^L_\xi < \aleph_1$ is much harder. In fact we shall prove the following result which is strictly stronger when ξ is a successor ordinal:

Main theorem. *For any recursive ordinal ξ:*

$$\aleph^L_\xi < \aleph_1 \implies \mathit{Lift}(\Pi^1_1, \Gamma_\xi)$$

The proof of this result is quite long and will be given in Chapter 6. This proof will make use of all the material of Chapters 1 and 2, including the double-tree representation of Borel sets. Once again the reader can check then that all the arguments used there are uniform in any given real parameter.

Moreover observe that the statement $\mathit{Lift}\,(\Pi^1_1, \Sigma^0_{1+\xi+1})$ is of the form "$\varphi \Rightarrow \psi$" where ψ a Σ_2-formula. Then by a standard absoluteness argument one gets the following more precise version which we will need for some of the applications:

THEOREM 4.4.4. *Fix $\alpha \in \omega^\omega$; let ξ be an ordinal which is countable in $L(\alpha)$, and assume that $\aleph_\xi^{L(\alpha)} < \aleph_1$. For any $\mathbf{\Gamma}_\xi$ set $Y \subset 2^\omega$ with $\mathbf{\Gamma}_\xi$-code in $L(\alpha)$, any $\mathbf{\Pi}^1_1$ set $Z \subset 2^\omega$ with $\mathbf{\Pi}^1_1$-code in $L(\alpha)$, any $\mathbf{\Delta}^1_1$ set $X_0 \subset 2^\omega \times 2^\omega$ with $\mathbf{\Delta}^1_1$-code in $L(\alpha)$, if any compact subset K of Y there exists $z \in Z$ such that $K \times \{z\} \subset X_0$, then Y has a continuous lifting in $X := X_0 \cap (2^\omega \times Z)$ with code in $L(\alpha)$.*

The notions of "$\mathbf{\Gamma}$-*code* of a given set A" used in this statement are to be taken in any reasonable sense. For example if $\mathbf{\Gamma} = \mathbf{\Gamma}_\xi$ one can consider any standard notion of Borel code. If $\mathbf{\Gamma} = \mathbf{\Pi}^1_1$ one can take for a $\mathbf{\Pi}^1_1$-code of A any sequential tree S on $2 \times \omega$ such that A is the complement of the projection of $\lceil S \rceil$. Finally if $\mathbf{\Gamma} = \mathbf{\Delta}^1_1$ one can take for a $\mathbf{\Delta}^1_1$-code of the set A a pair of $\mathbf{\Pi}^1_1$-codes for A and A^c.

CHAPTER 5

More Consequences and Reverse Results

Throughout this chapter $\Gamma \subset \Delta^1_1$ *denotes some lightface class of Borel sets and* $\mathbf{\Gamma} \subset \mathbf{\Delta}^1_1$ *its corresponding boldface class.*

Assuming the Main Theorem (see Section 4.4.3), we shall investigate the validity of the following statements discussed in the Introduction:

$\mathcal{L}ift(\Pi^1_1, \Gamma)$: *"For any* $X \subset 2^\omega \times 2^\omega$ *in* Π^1_1 *and any* $Y \subset 2^\omega$ *in* Γ*, if any compact subset of* Y *has a continuous lifting in* X *then* Y *has a continuous lifting in* X*"*

$\mathbb{A}(\Pi^1_1, \Gamma)$: *"For any* $X \subset 2^\omega \times 2^\omega$ *in* Π^1_1 *and any* $Y \subset 2^\omega$ *in* Γ*, if* $\pi_{|X}$ *is compact covering on* Y *then* $\pi_{|X}$ *is inductively perfect on* Y*"*

$\mathcal{N}_{\{\leq\}}(\Pi^1_1, \mathbf{\Gamma})$: *"Any* Π^1_1 *set admitting a* Π^1_1*-norm with all initial segments in* $\mathbf{\Gamma}$ *is in* $\mathbf{\Gamma}$*"*

$\mathcal{N}_{\{=\}}(\Pi^1_1, \mathbf{\Gamma})$: *"Any* Π^1_1 *set admitting a* Π^1_1*-norm with all constituents in* $\mathbf{\Gamma}$ *is in* $\mathbf{\Gamma}$*"*

We recall that if $\varphi : C \to \omega_1$ is a $\mathbf{\Pi}^1_1$-norm on the $\mathbf{\Pi}^1_1$ set C then by a φ-constituent we mean a set of the form $\{\varphi = \mu\}$, and by a φ-initial segment we mean a set of the form $\{\varphi \leq \mu\}$. Notice that the first two statements involve the lightface class Γ whereas the latest ones involve the boldface class $\mathbf{\Gamma}$.

We shall give in fact the exact logical strength of each of these statements for any class $\Gamma \subset \Delta^1_1$. For example we shall prove that for any recursive ordinal ξ if we set $\eta = 1 + \xi + 1$ then the following equivalences hold:

$$\aleph^L_\xi < \aleph_1 \iff \mathcal{L}ift(\Pi^1_1, \Sigma^0_\eta) \iff \mathbb{A}(\Pi^1_1, \Sigma^0_\eta) \iff \mathcal{N}_{\{\leq\}}(\Pi^1_1, \mathbf{\Pi}^0_\eta) \iff \mathcal{N}_{\{=\}}(\Pi^1_1, \mathbf{\Sigma}^0_\eta)$$

but if one replaces the Baire classes Σ^0_η, Π^0_η by their dual classes Π^0_η, Σ^0_η then these statements are no more equivalent and the exact strength of each of them depends on whether ξ is a limit or a successor ordinal.

The chapter is essentially constituted of two parts. In the first part (Section 5.1) we shall compare the various statements above, without any set theoretical assumption. Then applying Main Theorem we shall derive these statements from "$\aleph^L_\xi < \aleph_1$" (for some suitable ξ). In the second part (Section 5.2) we shall show that this large cardinal assumption is necessary.

The arguments developed in these two parts are totally independent each from the other and from the previous chapters. In particular the representation theorem of Borel sets is not needed here.

In the last section we shall study a generalization of $\mathcal{N}_{\{=\}}(\Pi^1_1, \mathbf{\Gamma})$ which can be stated as a Perfect Set Theorem for a class of equivalence relations, in the spirit of earlier results by Stern ([**25**], [**26**]).

5.1. Some boudedness principles

As explained in the Introduction the consideration of the statement $\mathcal{L}ift(\Pi^1_1, \Gamma)$ was originally motivated by the study of compact covering mappings. We also point out that by Theorem 3.1 of [**2**] the following general statement:

"For any mapping $f : X \to Y$ between two separable metrizable spaces with $X \in \Pi^1_1$ and $Y \in \Gamma$, if f is compact covering then f is inductively perfect."

is equivalent to $\mathbb{A}(\Pi^1_1, \Gamma)$. Moreover

PROPOSITION 5.1.1. *For any class $\Gamma \subset \Pi^1_1$:*

$$\mathcal{L}ift(\Pi^1_1, \Gamma) \implies \mathbb{A}(\Pi^1_1, \Gamma)$$

PROOF. Assume X is a Π^1_1 subset of $2^\omega \times 2^\omega$, $Y \subset 2^\omega$ is in Γ and $\pi_{|X}$ is compact-covering. Define a Π^1_1 set $X^* \subset 2^\omega \times \mathcal{K}(2^\omega)$ by

$$X^* := \{(y, L) \in 2^\omega \times \mathcal{K}(X) : y \in \pi(L)\}$$

Since $\pi_{|X}$ is compact-covering it is easy to check that for all compact $K \subset Y$ there exists a $L \in \mathcal{K}(X)$ such that $K \times \{L\} \subset X^*$. And since $\mathcal{K}(2^\omega)$ is recursively homeomorphic to 2^ω, it follows from $\mathcal{L}ift(\Pi^1_1, \Gamma)$ that there exists a continuous $f : Y \to \mathcal{K}(X)$ such that $y \in \pi(f(y))$ for all $y \in Y$. Then the set

$$H := \{(y, z) \in Y \times 2^\omega : (y, z) \in f(y)\}$$

is closed in $Y \times 2^\omega$, contained in X and satisfies $\pi(H) = Y$; hence $\pi_{|X}$ is inductively perfect. $\square$

We shall relate the statement $\mathbb{A}(\Pi^1_1, \Gamma)$ to a more explicit boundedness principle.

DEFINITION 5.1.2. We shall say that a $\mathbf{\Pi}^1_1$ set C has $\mathbf{\Gamma}$–approximations if for any $\mathbf{\Sigma}^1_1$ set $A \subset C$ there exists $B \in \mathbf{\Gamma}$ such that $A \subset B \subset C$.

In this section we shall study the following statement :

App$(\Pi^1_1, \mathbf{\Gamma})$: *"Any Π^1_1 set which admits $\mathbf{\Gamma}$-approximations is in $\mathbf{\Gamma}$"*

or more explicitly:

App$(\Pi^1_1, \mathbf{\Gamma})$: *"For any Π^1_1 set C, if for any $\mathbf{\Sigma}^1_1$ set $A \subset C$ there exists $B \in \mathbf{\Gamma}$ such that $A \subset B \subset C$, then C is in $\mathbf{\Gamma}$"*

Notice that this statement mixes lightface and boldface classes.

THEOREM 5.1.3. *If $\Gamma \subset \Delta^1_1$ is closed under unions with Σ^0_2 sets then:*

$$\mathbb{A}(\Pi^1_1, \Gamma) \implies App(\Pi^1_1, \check{\mathbf{\Gamma}})$$

REMARK 5.1.4. The restriction on Γ in the hypothesis of Theorem 5.1.3 is necessary: indeed if $\Gamma = \Pi^0_2$ (which is not closed under unions with Σ^0_2) then by [**3**], $\mathbb{A}(\Pi^1_1, \Gamma)$ is equivalent to "*Dom*"; while $App(\Pi^1_1, \mathbf{\Sigma}^0_2)$ implies that C_1 (the largest Π^1_1 thin set) is Borel hence countable, and this by classical properties of C_1 implies that "$\aleph_1^L < \aleph_1$", which cannot be derived from "*Dom*".

PROOF. For the proof of Theorem 5.1.3 we need a number of elementary results which are probably well known, but which we prove next for completeness.

Notations: Let $\mathbb{Q}$ denote the set of all eventually null sequences of 0 and 1, that is:

$$\mathbb{Q} := \{\alpha \in 2^\omega : \exists n : \forall p \geq n \quad \alpha(p) = 0\}$$

For any countable compact subset K of 2^ω, let $\mathrm{rk_c}(K)$ denote its Cantor rank; we recall that:

$$\mathrm{rk_c}(K) := \min\{\xi < \omega_1 : K^{(\xi)} = \emptyset\}$$

where $K^{(\xi)}$ is the ξth derivative of K.

For any space X we denote by $\mathscr{K}(X)$ the space of all non-empty compact subsets of X equipped with the Vietoris topology, and by $\mathscr{K}_\omega(X)$ the subspace of $\mathscr{K}(X)$ of all countable compact subsets of X.

We recall that $\mathscr{K}(2^\omega)$ is a compact space, and one can easily endow $\mathscr{K}(2^\omega)$ with a recursive presentation for which the sets $\{(K, s) \in \mathscr{K}(2^\omega) \times 2^{<\omega} : K \cap N_s \neq \emptyset\}$ and $\{(K, (s_i)_{1 \leq i \leq n}) \in \mathscr{K}(2^\omega) \times (2^{<\omega})^n : K \subset \bigcup_{1 \leq i \leq n} N_{s_i}\}$ are recursive, where N_s denote the basic clopen sets in 2^ω.

LEMMA 5.1.5. *Let K and L be countable compact subsets of 2^ω. If $\mathrm{rk_c}(K) < \mathrm{rk_c}(L)$ then there exists a continuous function f from 2^ω to itself such that $K \subset f(L)$.*

PROOF. The countable compact sets K and L are respectively homeomorphic to some countable ordinals $\eta + 1$ and $\zeta + 1$ each equipped with the order topology. Moreover $\eta < \omega^{\mathrm{rk_c}(K)+1}$ and $\zeta \geq \omega^{\mathrm{rk_c}(L)}$. Thus since $\mathrm{rk_c}(K) < \mathrm{rk_c}(L)$ we have $\eta < \zeta$. So K is homeomorphic to a closed subset K' of L. Then if $h : K' \to K$ is this homeomorphism and ρ a continuous retraction from 2^ω onto K', then $f := h \circ \rho$ defines a continuous mapping from 2^ω into itself such that $f(L) \supset h \circ \rho(K') = h(K') = K$. □

LEMMA 5.1.6. *There is a recursive mapping $\Phi : \mathscr{K}(2^\omega) \to \mathscr{K}(2^\omega)$ such that*

$$K \in \mathscr{K}(\mathbb{Q}) \iff \Phi(K) \in \mathscr{K}(\mathbb{Q}) \iff \Phi(K) \in \mathscr{K}_\omega(2^\omega)$$

PROOF. For α and β in 2^ω, write "$\alpha \leq \beta$" for "$\forall n,\ \alpha(n) \leq \beta(n)$", and define for $K \in \mathscr{K}(2^\omega)$:

$$\Phi(K) = \{\alpha \in 2^\omega : \exists \beta \in K \quad \alpha \leq \beta\}$$

Since $\Phi(K)$ is the first projection on 2^ω of the closed subset $\{(\alpha, \beta) : \alpha \leq \beta \text{ and } \beta \in K\}$ of $2^\omega \times 2^\omega$, it is compact. Since $\alpha \leq \beta$ and $\beta \in \mathbb{Q}$ imply $\alpha \in \mathbb{Q}$, we have

$$K \in \mathscr{K}(\mathbb{Q}) \implies \Phi(K) \in \mathscr{K}(\mathbb{Q}) \implies \Phi(K) \in \mathscr{K}_\omega(2^\omega)$$

And if $\beta_0 \in K \setminus \mathbb{Q}$, the set $\{\alpha : \alpha \leq \beta_0\}$ is homeomorphic to 2^ω, hence uncountable, and so is $\Phi(K) \supset \{\alpha : \alpha \leq \beta_0\}$. Thus

$$\Phi(K) \in \mathscr{K}_\omega(2^\omega) \implies K \in \mathscr{K}(\mathbb{Q})$$

Finally one easily checks that the graph of Φ is Π^0_1, and since the range space is a compact recursively presented space then Φ is recursive. □

LEMMA 5.1.7. *For any Π^1_1 subset Z of 2^ω, there exists a continuous recursive one-to-one mapping $\varphi : 2^\omega \to \mathscr{K}(2^\omega)$ such that :*

$$\alpha \in Z \iff \varphi(\alpha) \in \mathscr{K}(\mathbb{Q}) \iff \varphi(\alpha) \in \mathscr{K}_\omega(2^\omega)$$

PROOF. Since $\mathscr{K}(\mathbb{Q})$ is Π^1_1-complete, there exists a continuous recursive mapping φ_0 from 2^ω to $\mathscr{K}(2^\omega)$ reducing Z to $\mathscr{K}(\mathbb{Q})$; then $\varphi_1 := \Phi\circ\varphi_0$ satisfies

$$\alpha \in Z \iff \varphi_1(\alpha) \in \mathscr{K}(\mathbb{Q}) \iff \varphi_1(\alpha) \in \mathscr{K}_\omega(2^\omega)$$

But we have to construct a one-to-one such mapping. For this first define a mapping $\varphi_1' : 2^\omega \to \mathscr{K}(2^\omega)$ by $\varphi_1'(\alpha) = \{1^\frown x : x \in \varphi_1(\alpha)\}$; then clearly:

$$\varphi_1'(\alpha) \in \mathscr{K}(\mathbb{Q}) \iff \varphi_1(\alpha) \in \mathscr{K}(\mathbb{Q}) \text{ and } \varphi_1'(\alpha) \in \mathscr{K}_\omega(2^\omega) \iff \varphi_1(\alpha) \in \mathscr{K}_\omega(2^\omega)$$

For $n \in \omega$ let e_n denote the characteristic function on ω of $\{n+1\}$; and define $\varphi_2 : 2^\omega \to \mathscr{K}(\mathbb{Q})$ by $\varphi_2(\alpha) := \{\mathbf{0}\} \cup \{e_n : \alpha(n) = 1\}$. It is clear that φ_2 is continuous and one-to-one. Since $\beta(0) = 0$ for all $\beta \in \varphi_2(\alpha)$ and $\beta(0) = 1$ for all $\beta \in \varphi_1'(\alpha)$, it is easy to check that the mapping φ defined by $\varphi(\alpha) := \varphi_1'(\alpha) \cup \varphi_2(\alpha)$ is recursive, one-to-one and satisfies

$$\alpha \in Z \iff \varphi(\alpha) \in \mathscr{K}(\mathbb{Q}) \iff \varphi(\alpha) \in \mathscr{K}_\omega(2^\omega)$$

□

The proof of the next result that we omit follows directly from the definitions.

LEMMA 5.1.8. *Let $\mathbf{\Gamma}$ be a class closed under intersections with $\mathbf{\Pi}^0_2$ sets and let C be a $\mathbf{\Pi}^1_1$ set with $\mathbf{\Gamma}$-approximations.*

a) If C' is homeomorphic to C then C' has $\mathbf{\Gamma}$-approximations.

b) If P is a Polish space then $P \times C$ has $\mathbf{\Gamma}$-approximations.

c) If D is a relative $\mathbf{\Pi}^0_2$ subset of C then D has $\mathbf{\Gamma}$-approximations.

LEMMA 5.1.9. *Let $\mathbf{\Gamma}$ be a class closed under intersections with $\mathbf{\Pi}^0_2$ sets. If there exists a non-Borel Π^1_1 set $C \subset 2^\omega$ with $\mathbf{\Gamma}$-approximations, then for every Δ^1_1 subset Y of 2^ω there exists a Π^1_1 set $X \subset 2^\omega \times 2^\omega$ with $\mathbf{\Gamma}$-approximations such that:*

i) $\pi(X) = Y$

ii) For any compact subset K of Y, there exists $z \in 2^\omega$ such that $K \times \{z\} \subset X$

PROOF. By Lemma 5.1.7, C is homeomorphic to some *relatively* closed subset C' of $\mathscr{K}_\omega(2^\omega)$. So C' is a Π^1_1 non-Borel subset of $\mathscr{K}_\omega(2^\omega)$ with $\mathbf{\Gamma}$-approximations. Let $\mathscr{C}(2^\omega)$ denote the space of continuous functions from 2^ω to itself equipped with the uniform topology. Fix any natural recursive presentation of the Polish space $\mathscr{C}(2^\omega)$, and a recursive homeomorphism j from $\mathscr{C}(2^\omega) \times \mathscr{K}(2^\omega)$ onto some Π^0_2 subset $\hat{Z}$ of 2^ω, and define $Z := j(\mathscr{C}(2^\omega) \times C')$ which is a Π^1_1 subset of $\hat{Z}$. By Lemma 5.1.7 again there is a recursive one-to-one mapping $\varphi : 2^\omega \to \mathscr{K}(2^\omega)$ such that

$$y \in Y \iff \varphi(y) \in \mathscr{K}(\mathbb{Q}) \iff \varphi(y) \in \mathscr{K}_\omega(2^\omega)$$

Consider then

$$H := \{(y,z) \in 2^\omega \times 2^\omega : z \in \hat{Z} \text{ and } \varphi(y) \subset f(L) \text{ for } (f,L) := j^{-1}(z)\}$$

and

$$X := H \cap (2^\omega \times Z)$$

Clearly H is Π^0_2 and by Lemma 5.1.8 X is a Π^1_1 subset of $2^\omega \times 2^\omega$ with $\mathbf{\Gamma}$-approximations.

We now show that $\pi(X) \subset Y$. Let $(y,z) \in X$ and set $(f,L) := j^{-1}(z) \in \mathscr{C}(2^\omega) \times C'$. Since the compact set L is countable then so is $f(L)$, and since $\varphi(y) \subset f(L)$ then $\varphi(y) \in \mathscr{K}_\omega(2^\omega)$ hence $y \in Y$.

We claim that $\mathrm{rk_c}$ is not bounded below ω_1 on C'. Indeed, if there were some countable ordinal ξ such that $\mathrm{rk_c}(L) \leq \xi$ for all $L \in C'$, then the set $B_\xi := \{M \in \mathscr{K}_\omega(2^\omega) : \mathrm{rk_c}(M) \leq \xi\}$ would be Borel and since C' is relatively closed in $\mathscr{K}_\omega(2^\omega)$ we would have :

$$L \in C' \implies L \in C' \cap B_\xi \implies L \in \overline{C'} \cap B_\xi \implies L \in \overline{C'} \cap \mathscr{K}_\omega(2^\omega) \implies L \in C'$$

hence $C' = \overline{C'} \cap B_\xi$ would also be Borel, in contradiction with the hypothesis.

Hence for any compact subset K of Y, $\varphi(K)$ is a compact subset of $\mathscr{K}(\mathbb{Q})$, and $K^* := \bigcup\{\varphi(y) : y \in K\}$ is a compact subset of $\mathscr{K}(2^\omega)$ contained in $\mathbb{Q}$. Since K^* is countable, $\mathrm{rk_c}(K^*) < \omega_1$, and there is some $L \in C$ such that $\mathrm{rk_c}(K^*) < \mathrm{rk_c}(L)$. Then by Lemma 5.1.5 there is some $f \in \mathscr{C}(2^\omega)$ such that $K^* \subset f(L)$. Then $z := j(f, L) \in Z$ and for every $y \in K$ we have $\varphi(y) \subset K^* \subset f(L)$, hence $(y, z) \in H$. So $K \times \{z\} \subset X$, and the proof is complete. □

We recall that all classes considered are Lavrentieff classes (see 0.1).

<u>*Proof of Theorem 5.1.3*</u>:

Assume that $App(\Pi^1_1, \check{\mathbf{\Gamma}})$ fails; we shall prove that $\mathbb{A}(\Pi^1_1, \Gamma)$ fails. Let Y be a Γ-complete subset of 2^ω. Applying Lemma 5.1.9 above to the class $\check{\mathbf{\Gamma}}$, which is closed under intersection with $\mathbf{\Pi}^0_2$ sets, we get a Π^1_1 set $X \subset 2^\omega \times 2^\omega$ having $\check{\mathbf{\Gamma}}$-approximations such that $Y = \pi(X)$ and that every compact subset K of Y is the projection of some compact subset $K \times \{z\}$ of X. If $\mathbb{A}(\Gamma)$ held, $\pi_{|X}$ would be inductively perfect on Y and there would exist some closed subset F of $Y \times 2^\omega$ contained in X such that $\pi(F) = Y$. Then F would be a $\mathbf{\Sigma}^1_1$ subset of X, and there would exist some $\check{\mathbf{\Gamma}}$ set B such that $F \subset B \subset X$. It follows that $\pi_{|B}$ would be compact-covering (and even inductively perfect). And since $\mathbf{\Gamma} = D_\eta(\mathbf{\Sigma}^0_\xi)$ or $\mathbf{\Gamma} = \check{D}_\eta(\mathbf{\Sigma}^0_\xi)$ for $\eta \geq 1$ and $\xi \geq 2$, an easy extension of Theorem 8.6 in [**2**] would then imply that $Y \in \check{\mathbf{\Gamma}}$, in contradiction with the choice of Y. □

PROPOSITION 5.1.10. *For any class* $\mathbf{\Gamma} \subset \mathbf{\Delta}^1_1$*:*

$$App(\Pi^1_1, \mathbf{\Gamma}) \implies \mathscr{N}_{\{\leq\}}(\Pi^1_1, \mathbf{\Gamma})$$

Moreover if $\mathbf{\Gamma}$ *is an additive Baire class then:*

$$\mathscr{N}_{\{\leq\}}(\Pi^1_1, \mathbf{\Gamma}) \implies \mathscr{N}_{\{=\}}(\Pi^1_1, \mathbf{\Gamma})$$

PROOF. The first part follows from the classical boundedness Theorem for Π^1_1-norms. The second part is obvious since any initial segment is a countable union of constituents. □

Let recall the following particular classes considered in the Introduction and which will play a central role in our study:

if ξ is successor :
$\Sigma^*_\xi = \Sigma^0_{1+\xi+1}$, $\Pi^*_\xi = \Pi^0_{1+\xi+1}$, $\Gamma_\xi = D(\Sigma^0_{1+\xi+1})$, $\check{\Gamma}_\xi = \check{D}(\Sigma^0_{1+\xi+1})$

if ξ is limit :
$\Sigma^*_\xi = \Sigma^0_{1+\xi}$, $\Pi^*_\xi = \Pi^0_{1+\xi}$, $\Gamma_\xi = \Sigma^0_{1+\xi+1}$, $\check{\Gamma}_\xi = \Pi^0_{1+\xi+1}$

THEOREM 5.1.11. *For any recursive ordinal* ξ*:*

$$\aleph^L_\xi < \aleph_1 \implies \mathbb{A}(\Pi^1_1, \Gamma_\xi) \implies App(\Pi^1_1, \check{\mathbf{\Gamma}}_\xi) \implies \mathscr{N}_{\{\leq\}}(\Pi^1_1, \check{\mathbf{\Gamma}}_\xi) \implies \mathscr{N}_{\{=\}}(\Pi^1_1, \mathbf{\Sigma}^*_\xi)$$

PROOF. The first implication follows from Main Theorem and Proposition 5.1.1. For the second implication we distinguish two cases:

– If $\xi = 0$ then $\check{\mathbf{\Gamma}}_0 = \mathbf{\Pi}^0_2$ and $App(\Pi^1_1, \mathbf{\Pi}^0_2)$ is a well known ZFC result which follows from the classical Hurewicz Theorem; hence the implication holds.

– If $\xi > 0$ then Γ_ξ is closed under unions with Σ^0_2 sets, and the implication follows from Theorem 5.1.3.

The last two implications are particular instances of Proposition 5.1.10. □

Our next goal is to prove that in Theorem 5.1.11 the assumption $\aleph^L_\xi < \aleph_1$ is best possible. More precisely we shall prove that if any of the other statements holds for the dual class then necessarily $\aleph^L_{\xi+1} < \aleph_1$. From this one can derive that all the implications in Theorem 5.1.11 are actually equivalences. However notice that the last implication is no more valid for the dual classes since the dual Baire class $\mathbf{\Pi}^*_\xi$ is not additive and one cannot apply Proposition 5.1.10. So for the dual classes we only have

$$\mathbb{A}(\Pi^1_1, \check{\Gamma}_\xi) \implies App(\Pi^1_1, \mathbf{\Gamma}_\xi) \implies \mathscr{N}_{\{\le\}}(\Pi^1_1, \mathbf{\Gamma}_\xi)$$

and we will need to prove both implications

$$\begin{cases} \mathscr{N}_{\{\le\}}(\Pi^1_1, \mathbf{\Gamma}_\xi) \implies \aleph^L_{\xi+1} < \aleph_1 \\ \mathscr{N}_{\{=\}}(\Pi^1_1, \mathbf{\Pi}^*_\xi) \implies \aleph^L_{\xi+1} < \aleph_1 \end{cases}$$

5.2. Reverse results

As explained above we want to prove that if $\aleph^L_{\xi+1} \ge \aleph_1$ then:

(1) *there exists some Π^1_1-norm on a Π^1_1 non-Borel set with all constituents in $\mathbf{\Pi}^*_\xi$.*

(2) *there exists some Π^1_1-norm on a Π^1_1 non-Borel set with all initial segments in $\mathbf{\Gamma}_\xi$.*

In fact we shall prove the following stronger result:

THEOREM 5.2.1. *For any recursive ordinal $\xi \ge 0$, if $\aleph^L_{\xi+1} \ge \aleph_1$ then there exists a Π^1_1 non-Borel set W_ξ, and a Π^1_1-norm φ_ξ on W_ξ satisfying:*

(1) all φ_ξ-constituents are $\mathbf{\Pi}^*_\xi$

(2) all φ_ξ-initial segments are $\mathbf{\Gamma}_\xi$.

REMARK 5.2.2. The proof of Theorem 5.2.1 will rely on a construction due to R. L. Sami ([**23**]) and before we give precise definitions we shall present now this construction informally and indicate the new problems we have to solve.

Restated in our terminology Theorem II of [**23**] asserts that if $\aleph^L_{\xi+1} \ge \aleph_1$ then there exists an equivalence relation $\tilde{E}_\xi$ with Π^1_1 non-Borel domain W_ξ and all equivalence classes $\mathbf{\Pi}^*_\xi$ but which does not satisfy the Perfect Set Theorem (*i.e.* has uncountably many equivalence classes but does not admit a perfect set of pairwise non equivalent elements).

In fact the domain W_ξ of $\tilde{E}_\xi$ is defined as the space of all (possibly ill-founded) models M of "$KP + V = L$" with domain ω which interprete some fixed $a \in \omega$ as an ordinal of cardinality $\aleph_\xi$ and such that the restriction of the ϵ^M relation to a (viewed as a partial relation R^M on ω) is well-founded. It follows that the restriction of the ϵ^M to $\aleph_\xi$ (of M) is also a well-founded relation S^M. One can consider then: $\varphi_\xi(M)$ the order type of R^M, and $\psi_\xi(M)$ the order type of S^M. Then Sami considers the equivalence relation $\tilde{E}_\xi$ defined by:

$$M\,\tilde{E}_\xi\,N \quad\Longleftrightarrow\quad \varphi_\xi(M)=\varphi_\xi(N) \text{ and } \psi_\xi(M)=\psi_\xi(N)$$

Each $\tilde{E}_\xi$-equivalence class is of the form $\{M:\ \varphi_\xi(M)=\mu\}\cap\{M:\ \varphi_\xi(M)=\nu\}$; and to compute the complexity of such sets Sami makes a crucial use of the simple but fundamental fact that the ordinal $\varphi_\xi(M)$ can be coded by a subset of $\psi_\xi(M)$ (see 5.2.8).

To prove Theorem 5.2.1 we shall consider the same φ_ξ, which happens to be a Π^1_1-norm on W_ξ So we are reduced again to compute the complexity of the constituents $\{M:\ \varphi_\xi(M)=\mu\}$ and the initial segments $\{M:\ \varphi_\xi(M)\le\mu\}$.

Both computations are non trivial. First for the constituents because we now have to ignore the value of $\psi_\xi(M)$ which was crucial in Sami's computation. But also for the initial segments which are a priori countable unions of constituents. For example for $\xi=1$ the constituents will be $\mathbf{\Pi}^0_3$ but still we will have to show that the initial segments are in $D(\mathbf{\Sigma}^0_3)$.

Each of these two computations needs a new idea. We shall first improve Sami's computations by showing (for example for $\xi=1$) that the constituents are not merely $\mathbf{\Pi}^0_3$ sets but also $\mathbf{\Pi}^0_2$ *relatively* to W_ξ. This basic fact combined with some topological arguments will then enable us to prove that the countable union of such sets is not an arbitrary $\mathbf{\Sigma}^0_4$ set but is actually $D(\mathbf{\Sigma}^0_3)$ as desired.

5.2.3. Basic models (R.L. Sami) **.** In the sequel "*model*" will always mean a "model with domain ω for the standard language $\langle\in,=\rangle$". If M is such a model we denote by $\in^M$ the interpretation of the $\in$-relation.

For any $a\in\omega$ let $R^M(a)$ denote the binary relation on ω defined by the restriction of $\in^M$ to the set $\{b\in\omega : b\in^M a\}$. Notice that if $M\models$ "*a is an ordinal*" then $R^M(a)$ is a linear ordering, and define:

$$O^M(a)=\begin{cases}\text{the ordinal type of } R^M(a) & \text{if } R^M(a) \text{ is well-founded}\\ \infty & \text{if not}\end{cases}$$

The notation "$O^M(a)<\infty$" will stand for:

($M\models$ 'a is an ordinal") and ($R^M(a)$ is well-founded)

Finally set:

$$o(M)=\sup\{O^M(a);\quad O^M(a)<\infty\}$$

Let $\mathfrak{M}$ denote the set of all models. Sami equips $\mathfrak{M}$ with a structure of recursively presented Polish space, isomorphic to ω^ω, and for which given any formula θ the set $\{(M,\bar{a})\in\mathfrak{M}\times\omega^{<\omega}:\ M\models\theta(\bar{a})\}$ is Δ^0_1. In particular given any recursively enumerable theory T the set $\{M\in\mathfrak{M}:\ M\models T\}$ is Π^0_1.

Let T_0 be a theory formalizing "*KP + V=L + Axiom of infinity*". A basic model is an *ω-standard* model of T_0. If M is a basic model then O^M induces an isomorphism from $\langle\{a\in\omega : O^M(a)<\infty\},\in^M\rangle$ onto $\langle o(M),\in\rangle$, which extends uniquely to an isomorphism from $(L_M,\in^M)$ where $L_M:=\{x:\exists\, a,\ O^M(a)<\infty$ and $M\models x\in^M L_a\}$, onto $(L_{o(M)},\in)$.

5.2.4. Some Π^1_1 sets of basic models. Set:

$$\mathfrak{M}_0 := \{M \in \mathfrak{M} : \ M \text{ is a basic model}\}$$

It is not difficult to check that $\mathfrak{M}_0$ is a Π^0_2 subset of $\mathfrak{M}$. Also given any integer a we shall consider the sets:

$$\mathfrak{M}_0(a) = \{M \in \mathfrak{M}_0 : \ M \models \text{“}a \text{ is an ordinal”}\}$$

$$\mathfrak{M}^*_0(a) = \{M \in \mathfrak{M}_0(a) : \ O^M(a) < \infty\}$$

then clearly $\mathfrak{M}_0(a)$ is Δ^0_1 in $\mathfrak{M}_0$ hence is Π^0_2, and it is not difficult to see that $\mathfrak{M}^*_0(a)$ is Π^1_1.

As observed by Sami, if M is a basic model then $o(M) \geq \omega_1^{\mathrm{CK}}$; further, statements involving recursive ordinals can be expressed “inside M”. For example for all $\xi < \omega_1^{\mathrm{CK}}$ one can explicit a formula σ_ξ such that: $(M \models \sigma_\xi) \iff (\aleph_\xi$ *exists in* $M)$. In particular :

$$\mathfrak{M}_\xi := \{M \in \mathfrak{M}_0 : \ \aleph_\xi \ \text{ exists in } M\}$$

is a Δ^0_1 subset of $\mathfrak{M}_0$. If $M \in \mathfrak{M}_\xi$ and $\aleph_\xi$ is realized in M by $c^M_\xi \in \omega$, set:

$$\Omega^M_\xi := O^M(c^M_\xi)$$

(in [**23**] Ω^M_ξ is denoted by ω^M_ξ, which seems a bit confusing to us).

Given any integer a, Sami considers in [**23**] the Σ^1_1 equivalence relation $\hat{E}_\xi$ on $\mathfrak{M}_\xi$ defined by:

$$M \, \hat{E}_\xi \, N \iff \Omega^M_\xi = \Omega^N_\xi \text{ and } O^M(a) = O^N(a)$$

and studies the restriction E_ξ of $\hat{E}_\xi$ to some invariant Π^1_1 subset of $\mathfrak{M}_\xi$. He proves that E_ξ-equivalence classes are all $\mathbf{\Pi}^*_\xi$. He also proves that E_ξ is always thin, but has uncountably many equivalence classes whenever $\aleph^L_{\xi+1}$ is uncountable. We shall not explicit here the domain of E_ξ which is slightly technical. But in next paragraph we shall consider some other, quite closely related but simpler, Π^1_1 subsets of $\mathfrak{M}_\xi$.

For any integer a consider first the Δ^0_1 subsets of $\mathfrak{M}_\xi$ defined by:

$$\mathfrak{M}_{\xi,a} := \{M \in \mathfrak{M}_\xi : \ M \models \text{“}a \text{ is an ordinal and } \mathrm{card}(a) = \aleph_\xi\text{”}\}$$

Finally define:

$$\mathfrak{M}^*_{\xi,a} := \{M \in \mathfrak{M}_{\xi,a} : \ O^M(a) < \infty\}$$

and for $\mu < \aleph_1$:

$$\mathfrak{M}^\mu_{\xi,a} := \{M \in \mathfrak{M}_{\xi,a} : \ O^M(a) = \mu\}$$

Notice that since all models are ω-standard, if $M \in \mathfrak{M}^*_{\xi,a}$ we necessarily have $O^M(a) \geq \omega$. But for some inductive arguments we will need $O^M(a)$ to cover the whole ordinal spectrum. For this we extend the previous definitions and set by convention:

$$\mathfrak{M}_{-1,a} = \mathfrak{M}^*_{-1,a} := \{M \in \mathfrak{M}_0 : \ O^M(a) \ \text{ is finite }\}$$

and for all $k \in \omega$:

$$\mathfrak{M}^k_{-1,a} := \{M \in \mathfrak{M}_0 : \ O^M(a) = k\}$$

However whenever we will speak about $\mathfrak{M}^\mu_{\xi,a}$ with no other precision it will be understood that ξ is an ordinal, so $\xi \geq 0$.

LEMMA 5.2.5. *$\mathfrak{M}^*_{\xi,a}$ is Π^1_1 and the mapping $\psi_{\xi,a} : M \mapsto O^M(a)$ defines a Π^1_1-norm on $\mathfrak{M}^*_{\xi,a}$. Moreover if $\aleph^L_{\xi+1} = \aleph_1$ then $\mathfrak{M}^*_{\xi,a}$ is non-Borel.*

PROOF. For the first part notice that $\psi_{\xi,a}$ is just the composition of the Δ^1_1 mapping $M \mapsto R^M(a)$ considered in 5.2.3, from $\mathfrak{M}_{\xi,a}$ to $\mathscr{P}(\omega \times \omega)$, with the canonical norm on WO. The last part follows from the proof of ([**23**], Claim 3.3.4) □

Next two notions will play a central role in the proof of Theorem 5.2.1.

5.2.6. Relative descriptive complexity. Up to now, and as explicitly stated in the Introduction, the descriptive complexity of a given set A was always considered relatively to some (arbitrary) *compact* (or at least Polish) space X in which A is embedded; and the compactness requirement was present precisely because in this context complexity computations are absolute and do not depend on the surrounding space X. But of course one can always compute the complexity of A in any other (non necessarily compact or Polish) space; the result will in general be simpler than the absolute computation in compact spaces, which is in fact the upper bound of these complexities. For example A is always $\mathbf{\Delta}^0_1$ relative to the space A, but of course other intermediate complexities might occur. For practical reasons we shall slightly enlarge the notion of relative complexity.

<u>*Terminology*</u>: Let $\mathbf{\Gamma}$ be a "boldface" descriptive class. Given any two subsets A and B of a separable metrizable space X, we shall say that *A is $\mathbf{\Gamma}$ <u>in</u> B* if $A \cap B$ is a $\mathbf{\Gamma}$-subset of the topological space B.

When we will speak about the complexity of a set A without any reference to a particular space, we shall mean as before the *absolute* complexity of A, that is the complexity of A in some (any) compact metric space.

LEMMA 5.2.7. *Suppose that $M, N \in \mathfrak{M}_\xi$ satisfy both that "a is an ordinal and* $\text{card}(a) \geq \aleph_\xi$*". If $O^M(a) = O^N(a)$ then $\Omega^M_\xi = \Omega^N_\xi$.*

PROOF. We fix a and prove the statement by induction on ξ. For $\xi = 0$ this is obvious since any basic model M satisfies $\Omega^M_\xi = \omega$.

Suppose that the statement is true for all $\eta < \xi$.

– If ξ is limit notice that $\mathfrak{M}_\xi = \bigcap_{\eta<\xi} \mathfrak{M}_\eta$ and $\Omega^M_\xi = \sup_{\eta<\xi} \Omega^M_\eta$ for all $M \in \mathfrak{M}_\xi$. Hence the statement for ξ follows straightforward from the induction hypothesis.

– If $\xi = \eta + 1$ is successor we argue by contradiction; so suppose that $\Omega^M_\xi < \Omega^N_\xi$. Then again by the induction hypothesis $\Omega^M_\eta = \Omega^N_\eta < \Omega^M_{\eta+1} < \Omega^N_{\eta+1} \leq O^M(a) = O^N(a) < \infty$. Since $L_{o(N)} \models$ "$\aleph_\eta = \underline{\Omega}^N_\eta$ and $\aleph_{\eta+1} = \underline{\Omega}^N_{\eta+1}$" then $L_{o(N)} \models$ "$\text{card}(\underline{\Omega}^M_{\eta+1}) = \aleph_\eta$", hence by Gödel's condensation lemma for any $\lambda \geq \Omega^N_{\eta+1}$ we also have $L_\lambda \models$ "$\text{card}(\underline{\Omega}^M_{\eta+1}) = \aleph_\eta$" and applying this observation to $\lambda = \Omega^M_{\eta+1}$ we get a contradiction. □

5.2.8. L-codes of ordinals (Sami). This notion is the basic tool of the computations in [**23**]. Fix for any infinite ordinal λ a bijection $j_\lambda : \lambda \to \lambda \times \lambda$, definable in L uniformly in λ. If $\mu \geq \lambda$ is also an ordinal, we shall say that $A \subset \lambda$ is the <u>L-code</u> of μ on λ, if A is the $<_L$-least element in L such that $j'_\lambda(A) \subset \lambda \times \lambda$ is a well-ordering on λ of type μ (notice that this necessitates that $L \models \text{card}(\mu) \leq \lambda$).

For any recursive ordinal ξ one can clearly explicit a formula $\theta_\xi(x,y)$ which formalizes in some standard way the statement:

"*(y is an ordinal of cardinality $\aleph_\xi$) and (x is the L-code of y on $\aleph_\xi$)*"

We shall now combine one of Sami's main arguments with Lemma 5.2.7 above.

LEMMA 5.2.9. *For all $\mu < \omega_1$ there exists $C \subset \mu$ such that for any basic model M, if $M \models \theta_\xi(c,a)$ then:*

$$M \in \mathfrak{M}^\mu_{\xi,a} \iff C = \{O^M(b) :\ b \in^M c\}$$

PROOF. If $\mathfrak{M}^\mu_{\xi,a} = \emptyset$ take $C = \emptyset$; then the equivalence holds since on both sides the statements are false (notice that any finite integer k is of the form $O^M(b)$).

If $\mathfrak{M}^\mu_{\xi,a} \neq \emptyset$ then by Lemma 5.2.7 there exists a uniquely determined ordinal $\lambda \leq \mu$ such that for any $M \in \mathfrak{M}^\mu_{\xi,a}$ we have $\Omega^M_\xi = \lambda$. Let $C \subset \lambda$ be the L-code of μ on λ.

If $M \in \mathfrak{M}^\mu_{\xi,a}$ then $L_{o(M)} \models$ "card$(\underline{\mu}) = \aleph_\xi = \underline{\lambda} \leq \underline{\mu}$". Notice that by absoluteness of well-foundedness, C is also the L-code of μ on λ, computed in $L_{o(M)}$. Since $O^M(a) < \infty$ and c is definable from a then $c \in L_M$ (see 5.2.3) and by $(\in^M, \in)$-isomorphism we have: $C = \{O^M(b)\,;\ b \in^M c\}$. Conversely if the previous equality holds, then because of the $(\in^M, \in)$-isomorphism, $O^M(a)$ is just the height of the well-founded relation coded by C and which is precisely μ. □

LEMMA 5.2.10. *For any recursive ordinal ξ:*

a) If ξ is limit then $\mathfrak{M}^\mu_{\xi,a}$ is $\mathbf{\Pi}^0_{1+\xi}$ in $\mathfrak{M}_0$.

b) If ξ is successor then $\mathfrak{M}^\mu_{\xi,a}$ is $\mathbf{\Pi}^0_{1+\xi+1}$ in $\mathfrak{M}_0$.

PROOF. Notice first that for any a and c the set $\mathfrak{P}_{a,c} := \{M \in \mathfrak{M}_0 : M \models \theta_\xi(a,c)\}$ is Δ^0_1 in $\mathfrak{M}_0$. Moreover for any fixed a the family $(\mathfrak{P}_{a,c})_{c\in\omega}$ defines a clopen partition of $\mathfrak{M}_0$, and the complexity of any subset of $\mathfrak{M}_0$ is the supremum of the complexities of its traces on each $\mathfrak{P}_{a,c}$.

So given a, ξ, μ let C be as in Lemma 5.2.9. Then for any c and any $M \in \mathfrak{P}_{a,c}$:

$$(\star)\quad M \in \mathfrak{M}^\mu_{\xi,a} \iff \begin{cases} \forall \nu \in C\,, \exists \eta < \xi\,, \exists b \in \omega\,, M \models b \in c \text{ and } M \in \mathfrak{M}^\nu_{\eta,b} \\ \text{and} \\ \forall b \in \omega\,, M \not\models b \in c \text{ or } \exists \eta < \xi\,, \exists \nu \in C\,, M \in \mathfrak{M}^\nu_{\eta,b} \end{cases}$$

where η is allowed to take the value "-1".

We now prove the lemma (for all a and all μ) by induction on ξ.

For $\xi = 0$ the only possible value for η in $(\star)$ is -1, and since for all $k \in \omega$, $\mathfrak{M}^k_{-1,b} = \{M \in \mathfrak{M}_0 :\ M \models \text{"card}(b) = \underline{k}\text{"}\}$ is Δ^0_1 in $\mathfrak{M}_0$, then $\mathfrak{M}^\mu_{\xi,a}$ is Π^0_1 in $\mathfrak{M}_0$.

Let $\xi > 0$ and suppose that the lemma is true for all $\zeta < \xi$. Then it is clear from the formula on right hand side in $(\star)$ that $\mathfrak{M}^\mu_{\xi,a}$ is a countable intersection of sets from $\bigcup_{\eta<\xi} \mathbf{\Pi}^0_\eta$. Hence $\mathfrak{M}^\mu_{\xi,a}$ is $\mathbf{\Pi}^0_{1+\xi+1}$; and if moreover ξ is limit then $\mathfrak{M}^\mu_{\xi,a}$ is $\mathbf{\Pi}^0_{1+\xi}$. □

In the limit case next lemma is weaker than Lemma 5.2.10, but it is for the successor case that we will need this result later on.

LEMMA 5.2.11. *For any recursive ordinal ξ, $\mathfrak{M}^\mu_{\xi,a}$ is $\mathbf{\Pi}^0_{1+\xi}$ in $\mathfrak{M}^*_0(a)$.*

PROOF. Again given a, ξ, μ let C be as in Lemma 5.2.9. Then for any $M \in \mathfrak{M}_0^*(a) \cap \mathfrak{P}_{a,c}$:

$$(\star\star) \qquad M \in \mathfrak{M}_{\xi,a}^{\mu} \iff \begin{cases} \forall b \in \omega, \forall \nu < \lambda, \forall \eta < \xi \\ [M \in \mathfrak{M}_{\eta,b}^{\nu} \Rightarrow (M \models b \in c) \Leftrightarrow (\nu \in C)] \end{cases}$$

We also prove the lemma (for all a and all μ) by induction on ξ. For the limit case (in particular for $\xi = 0$) the result follows from Lemma 5.2.10.

Let ξ be a successor ordinal, and suppose that the lemma is true for all $\zeta < \xi$. Then by $(\star\star)$ $\mathfrak{M}_{\xi,a}^{\mu}$ is a countable intersection of sets which are either Δ_1^0 in $\mathfrak{M}_0$, or of the form $\mathfrak{M}_0 \setminus \mathfrak{M}_{\eta,b}^{\nu}$ for some b such that $M \models$ "$b \in c$ and $\operatorname{card}(b) = \aleph_\eta$" for some $\eta < \xi$, but since $O^M(a) < \infty$, then for such a model M:

$$M \in \mathfrak{M}_0 \setminus \mathfrak{M}_{\eta,b}^{\nu} \iff M \in \mathfrak{M}_0^*(a) \setminus \mathfrak{M}_{\eta,b}^{\nu} \iff M \in \mathfrak{M}_0^*(b) \setminus \mathfrak{M}_{\eta,b}^{\nu}$$

and by the induction hypothesis $\mathfrak{M}_0^*(b) \setminus \mathfrak{M}_{\eta,b}^{\nu}$ is $\mathbf{\Sigma}_{1+\eta}^0$ in $\mathfrak{M}_0^*(b)$. It follows that the set $\{M : M \models \text{"}b \in c \text{ and } \operatorname{card}(b) = \aleph_\eta\text{"}\} \cap \left(\mathfrak{M}_0^*(a) \setminus \mathfrak{M}_{\eta,b}^{\nu}\right)$ is $\mathbf{\Sigma}_{1+\eta}^0$ in $\mathfrak{M}_0^*(a)$. One easily derives from these observations that the set $\mathfrak{M}_{\xi,a}^{\mu}$ is $\mathbf{\Pi}_{1+\xi}^0$ in $\mathfrak{M}_0^*(a)$. □

LEMMA 5.2.12. *A subset A of a separable metrizable space X is $D(\mathbf{\Sigma}_{\xi+1}^0)$ in X if and only if $A = \bigcup_{j\in\omega} A_j$ is the union of a countable family $(A_j)_{j\in\omega}$ where each A_j is simultaneously a $\mathbf{\Pi}_{\xi+1}^0$ in X and $\mathbf{\Pi}_\xi^0$ in A.*

PROOF. If A is $D(\mathbf{\Sigma}_{\xi+1}^0)$ we can find two countable families $(B_j)_{j\in\omega}$ and $(C_j)_{j\in\omega}$ of $\mathbf{\Pi}_\xi^0$ sets such that $A = \bigcup_{j\in\omega} B_j \setminus \bigcup_{j\in\omega} C_j$. Then $C' = X \setminus \bigcup_{j\in\omega} C_j$ is $\mathbf{\Pi}_{\xi+1}^0$ so $A_j := B_j \cap C'$ is also $\mathbf{\Pi}_{\xi+1}^0$, and since $A_j \cap A = B_j \cap A$ then each A_j is relatively $\mathbf{\Pi}_\xi^0$ in A; finally we clearly have $A = \bigcup_{j\in\omega} A_j$.

Conversely if $A = \bigcup_{j\in\omega} A_j$ with each A_j $\mathbf{\Pi}_{\xi+1}^0$ and a relative $\mathbf{\Pi}_\xi^0$ subset of A, then there exists some $\mathbf{\Pi}_\xi^0$ subsets $B_j \subset X$ such that $A_j = B_j \cap A$. Then for each j, $(B_j \setminus A_j)^c \supset A$, hence

$$\left(\bigcup_{j\in\omega} B_j\right) \setminus \left(\bigcup_{j\in\omega} (B_j \setminus A_j)\right) \subset A = \left(\bigcup_{j\in\omega} B_j\right) \cap A \subset \left(\bigcup_{j\in\omega} B_j\right) \cap \left(\bigcup_{j\in\omega} (B_j \setminus A_j)\right)^c$$

Since $B_j \in \mathbf{\Pi}_\xi^0$ and $A_j \in \mathbf{\Pi}_{\xi+1}^0$, then $(B_j \setminus A_j) \in \mathbf{\Sigma}_{\xi+1}^0$, then $\bigcup_j B_j$ and $\bigcup_j (B_j \setminus A_j)$ are both $\mathbf{\Sigma}_{\xi+1}^0$, hence

$$A = \left(\bigcup_{j\in\omega} B_j\right) \setminus \left(\bigcup_{j\in\omega} (B_j \setminus A_j)\right)$$

is $D(\mathbf{\Sigma}_{\xi+1}^0)$. □

<u>*Proof of Theorem 5.2.1*</u>:

Since $\aleph_{\xi+1}^L \geq \aleph_1$ there exists $\eta \leq \xi$ such that $\aleph_{\eta+1}^L = \aleph_1$. Fix then any integer a, set $W_\xi := \mathfrak{M}_{\eta,a}^*$ and let φ_ξ denote the restriction to W_ξ of the canonical norm $M \mapsto O^M(a)$ on $\mathfrak{M}^*(a)$. By Lemma 5.2.5, W_ξ is non-Borel, and conditions (1) and (2) follow then from Lemma 5.2.10, Lemma 5.2.11 and Lemma 5.2.12. □

5.3. Conclusion

THEOREM 5.3.1. *Let ξ be any recursive ordinal. Then for any class Γ such that $\bigcup_{\eta<\xi}\Pi^*_\eta \subset \Gamma \subset \Sigma^*_\xi$*

$$\aleph^L_\xi < \aleph_1 \iff \mathscr{N}_{\{=\}}(\mathit{\Pi}^1_1, \mathbf{\Gamma})$$

PROOF. The implication $\aleph^L_\xi < \aleph_1 \Rightarrow \mathscr{N}_{\{=\}}(\mathit{\Pi}^1_1, \mathbf{\Sigma}^*_\xi)$ follows from Theorem 5.1.11 and we now prove the converse implication by induction.

For $\xi = 0$ there is nothing to prove. For the induction step we distinguish two cases:

– If $\xi = \eta + 1$ is a successor ordinal then, since $\Gamma \supset \Pi^*_\eta$, it follows from the hypothesis that $\mathscr{N}_{\{=\}}(\mathit{\Pi}^1_1, \mathbf{\Pi}^*_\eta)$ holds; hence by Theorem 5.2.1, $\aleph^L_{\eta+1} = \aleph^L_\xi < \aleph_1$

– If $\xi > 0$ is a limit ordinal then it follows from the hypothesis that for all $\eta < \xi$, $\mathscr{N}_{\{=\}}(\mathit{\Pi}^1_1, \mathbf{\Pi}^*_\eta)$ holds hence by the induction hypthesis, $\aleph^L_{\eta+1} < \aleph_1$; so $\aleph^L_\xi < \aleph_1$. □

Following exactly the same scheme one proves:

THEOREM 5.3.2. *Let ξ be any recursive ordinal. Then for any class Γ such that $\bigcup_{\eta<\xi}\check{\Gamma}_\eta \subset \Gamma \subset \Gamma_\xi$ and closed under unions with $\mathit{\Sigma}^0_2$ sets, each of the following statements is equivalent to $\aleph^L_\xi < \aleph_1$:*

$$\mathit{Lift}(\mathit{\Pi}^1_1, \Gamma) \; ; \; \mathbb{A}(\mathit{\Pi}^1_1, \Gamma) \; ; \; \mathit{App}(\mathit{\Pi}^1_1, \check{\mathbf{\Gamma}}) \; ; \; \mathscr{N}_{\{\leq\}}(\mathit{\Pi}^1_1, \check{\mathbf{\Gamma}})$$

PROOF. By Main Theorem, Proposition 5.1.1, Theorem 5.1.3 and Proposition 5.1.10 we have:

$$\aleph^L_\xi < \aleph_1 \Longrightarrow \mathit{Lift}(\mathit{\Pi}^1_1, \Gamma_\xi) \Longrightarrow \mathit{Lift}(\mathit{\Pi}^1_1, \Gamma)$$
$$\Longrightarrow \mathbb{A}(\mathit{\Pi}^1_1, \Gamma) \Longrightarrow \mathit{App}(\mathit{\Pi}^1_1, \check{\mathbf{\Gamma}}) \Longrightarrow \mathscr{N}_{\{\leq\}}(\mathit{\Pi}^1_1, \check{\mathbf{\Gamma}})$$

The proof of the implication $\mathscr{N}_{\{\leq\}}(\mathit{\Pi}^1_1, \check{\mathbf{\Gamma}}) \Rightarrow \aleph^L_\xi < \aleph_1$ is then achieved by following the same scheme as in the proof of Theorem 5.3.1. □

THEOREM 5.3.3. *Let ξ be a recursive successor ordinal. Then each of the following statements is equivalent to $\aleph^L_\xi < \aleph_1$:*

$$\mathit{Lift}(\mathit{\Pi}^1_1, \mathit{\Sigma}^0_{1+\xi+1}) \; ; \; \mathbb{A}(\mathit{\Pi}^1_1, \mathit{\Sigma}^0_{1+\xi+1}) \; ; \; \mathit{App}(\mathit{\Pi}^1_1, \mathbf{\Pi}^0_{1+\xi+1}) \; ; \; \mathscr{N}_{\{\leq\}}(\mathit{\Pi}^1_1, \mathbf{\Pi}^0_{1+\xi+1})$$
$$\mathit{Lift}(\mathit{\Pi}^1_1, \mathit{\Pi}^0_{1+\xi+1}) \; ; \; \mathbb{A}(\mathit{\Pi}^1_1, \mathit{\Pi}^0_{1+\xi+1}) \; ; \; \mathit{App}(\mathit{\Pi}^1_1, \mathbf{\Sigma}^0_{1+\xi+1}) \; ; \; \mathscr{N}_{\{\leq\}}(\mathit{\Pi}^1_1, \mathbf{\Sigma}^0_{1+\xi+1})$$

PROOF. Notice that since $\xi > 0$ both classes $\mathit{\Sigma}^0_{1+\xi+1}$ and $\mathit{\Pi}^0_{1+\xi+1}$ are closed under unions with $\mathit{\Sigma}^0_2$ sets. Moreover since ξ is a successor ordinal then $\Gamma_\xi = D(\mathit{\Sigma}^0_{1+\xi+1})$ hence $\bigcup_{\eta<\xi}\check{\Gamma}_\eta \subset \mathit{\Sigma}^0_{1+\xi+1} \subset \Gamma_\xi$ and $\bigcup_{\eta<\xi}\check{\Gamma}_\eta \subset \mathit{\Pi}^0_{1+\xi+1} \subset \Gamma_\xi$. □

COROLLARY 5.3.4. *Let ξ be a recursive limit ordinal. Then each of the following statements is equivalent to $\aleph^L_\xi < \aleph_1$:*

$$\mathit{Lift}(\mathit{\Pi}^1_1, \mathit{\Sigma}^0_{1+\xi+1}) \; ; \; \mathbb{A}(\mathit{\Pi}^1_1, \mathit{\Sigma}^0_{1+\xi+1}) \; ; \; \mathit{App}(\mathit{\Pi}^1_1, \mathbf{\Pi}^0_{1+\xi+1}) \; ; \; \mathscr{N}_{\{\leq\}}(\mathit{\Pi}^1_1, \mathbf{\Pi}^0_{1+\xi+1}$$

PROOF. For all ξ the class $\mathit{\Sigma}^0_{1+\xi+1}$ is closed under unions with $\mathit{\Sigma}^0_2$ sets. Since ξ is a limit ordinal then $\Gamma_\xi = \mathit{\Sigma}^0_{1+\xi+1}$ hence $\bigcup_{\eta<\xi}\check{\Gamma}_\eta \subset \mathit{\Sigma}^0_{1+\xi+1} \subset \Gamma_\xi$. □

COROLLARY 5.3.5. *Let ξ be a recursive infinite limit ordinal. Then each of the following statements is equivalent to $\aleph^L_{\xi+1} < \aleph_1$:*

$$\mathit{Lift}(\Pi^1_1, \Pi^0_{1+\xi+1}) \ ; \ \mathbb{A}(\Pi^1_1, \Pi^0_{1+\xi+1}) \ ; \ \mathit{App}(\Pi^1_1, \mathbf{\Sigma}^0_{1+\xi+1}) \ ; \ \mathscr{N}_{\{\leq\}}(\Pi^1_1, \mathbf{\Sigma}^0_{1+\xi+1})$$

PROOF. Again since $\xi > 0$ the class $\Pi^0_{1+\xi+1}$ is closed under unions with Σ^0_2 sets. Moreover $\Gamma_\xi = \Sigma^0_{1+\xi+1}$ and $\Gamma_{\xi+1} = D(\Sigma^0_{1+\xi+2})$ hence $\check{\Gamma}_\xi \subset \Pi^0_{1+\xi+1} \subset \Gamma_{\xi+1}$. □

We recall (see 4.4.3) that $\mathit{Lift}(\Pi^1_1, \Pi^0_2)$ and $\mathbb{A}(\Pi^1_1, \Pi^0_2)$ are equivalent to *Dom*; but $\mathit{App}(\Pi^1_1, \mathbf{\Sigma}^0_2)$ and $\mathscr{N}_{\{\leq\}}(\Pi^1_1, \mathbf{\Sigma}^0_2)$ are equivalent to $\aleph^L_1 < \aleph_1$.

We now come back to the initial Problem I considered in the Introduction. For simplicity we restrict the statement to projection mappings (see 0.2); for more general mappings one has to impose that they are Δ^1_1.

THEOREM 5.3.6. *For any recursive ordinal ξ the following are equivalent:*

(i) $\aleph^L_\xi < \aleph_1$

(ii) Any compact covering projection from a Π^1_1 set onto a Γ_ξ set is inductively perfect.

(iii) Any compact covering projection from a Γ_ξ set onto a Γ_ξ set is inductively perfect.

(iv) Any compact covering projection from a Γ_ξ set is inductively perfect.

PROOF. First observe that *(ii)* is precisely $\mathbb{A}(\Pi^1_1, \Gamma_\xi)$ so by Theorem 5.3.2, *(i)* $\Leftrightarrow$ *(ii)*.

The implication *(ii)* $\Rightarrow$ *(iii)* is trivial. The converse follows from Ostrovsky's result ([**20**]) for $\xi = 0$, and for $\xi > 0$ from the arguments of ([**3**] Theorem 6.5) since in this case $\check{D}(\Sigma^0_2) \subset \Gamma_\xi \subset \Delta^1_1$.

Hence *(i)* $\Leftrightarrow$ *(ii)* $\Leftrightarrow$ *(iii)*. Moreover *(iv)* $\Rightarrow$ *(iii)* is trivial and to finish we now prove *(i)* $\Rightarrow$ *(iv)*.

So assume that $\aleph^L_\xi < \aleph_1$ and suppose that $\pi_X : X \to \pi(X)$ is a compact covering projection with $X \in \Gamma_\xi$. By Remark 4.3.2 a), $Y := \pi(X)$ is in the (boldface) class $\mathbf{\Gamma}_\xi$, and by Remark 4.3.2 b) Y has a Borel code in L. Hence by a combined application of Theorem 4.4.4 and Proposition 5.1.1 one can conclude that π_X is inductively perfect. □

5.4. Appendix: A Perfect Set Theorem for a Class of Equivalence Relations

For simplicity we restrict our study in this section to relations whose domain is a subset of ω^ω, but all results extend to any recursively presented Polish space. To state simply the main results of this section, we shall use the following terminology.

DEFINITION 5.4.1. We shall say that an equivalence relation E on a space X is thin if there exists no perfect set in X of pairwise non-equivalent elements.

We shall say that E satisfies the Perfect Set Theorem if E is either thin or has at most countably many equivalence classes.

Thus Silver's classical Theorem asserts that any $\mathbf{\Pi}^1_1$ equivalence relation satisfies the Perfect Set Theorem, and it is well known that this is false for arbitrary $\mathbf{\Sigma}^1_1$ equivalence relations. But in [**26**] Stern proved the following result:

THEOREM 5.4.2. *(Stern) Let ξ be an ordinal which is countable in L. If $\aleph_\xi^L < \aleph_1$ then any Σ_1^1 equivalence relation with all equivalence classes in $\mathbf{\Sigma}^0_{1+\xi+2}$ satisfies the Perfect Set Theorem.*

We point out that when $\xi = 0$, the lightface feature of the hypothesis is irrelevant and in this case Theorem 5.4.2 holds for any $\mathbf{\Sigma}_1^1$ relation with equivalence classes in $\mathbf{\Sigma}^0_3$ ([**26**] Theorem 8.1) .

5.4.3. The equivalence relations E_φ. If φ is any $\mathbf{\Pi}_1^1$ -norm on a $\mathbf{\Pi}_1^1$ set C, one can define an equivalence relation E_φ on C by:

$$x\, E_\varphi\, y \iff \varphi(x) = \varphi(y)$$

Since a perfect set cannot admit a Borel well-ordering, the equivalence relation E_φ is always thin. In particular one can reformulate $\mathscr{N}_{\{=\}}(\Pi_1^1, \mathbf{\Gamma})$ as follows:

$$\mathscr{N}_{\{=\}}(\Pi_1^1, \mathbf{\Gamma}) \iff \begin{cases} \textit{For any } \Pi_1^1\textit{-norm } \varphi, \textit{ if all } E_\varphi\textit{-equivalence classes} \\ \textit{are in } \mathbf{\Gamma} \textit{ then } E_\varphi \textit{ satisfies the Perfect Set Theorem} \end{cases}$$

Observe that unless C is Borel, the equivalence relation E_φ is not $\mathbf{\Sigma}_1^1$ and Stern's Theorem does not apply in this case and our goal is to relax the condition "*E is Σ_1^1*" to cover the equivalence relations E_φ. For this we introduce the following notion.

DEFINITION 5.4.4. Given a class Λ, we shall say that an equivalence relation E with domain A is almost in Λ if there exists an equivalence relation $\hat{E}$ in Λ such that A is an $\hat{E}$-invariant subset of the domain of $\hat{E}$, and E is the restriction of $\hat{E}$ to A.

EXAMPLE 5.4.5. If φ is a $\mathbf{\Pi}_1^1$ -norm on a $\mathbf{\Pi}_1^1$ set C, the canonical equivalence relation E_φ on C is almost $\mathbf{\Sigma}_1^1$. To see this embed C in any $\mathbf{\Sigma}_1^1$ space $\hat{C}$ and define for $x, y \in \hat{C}$:

$$x\, \hat{E}_\varphi\, y \iff x, y \in \hat{C} \setminus C \text{ or } x\, E_\varphi\, y$$

since φ is a $\mathbf{\Pi}_1^1$-norm the equivalence relation $\hat{E}_\varphi$ is $\mathbf{\Sigma}_1^1$; moreover the domain C of E_φ (which in this case is $\mathbf{\Pi}_1^1$) is clearly $\hat{E}_\varphi$-invariant. Moreover, unless C is Borel, E_φ is not $\mathbf{\Sigma}_1^1$.

By the same arguments one sees that if C is Π_1^1 and φ is a Π_1^1 -norm then E_φ is almost Σ_1^1 with Π_1^1 domain.

Notice that in this example, C is not merely $\hat{E}_\varphi$-invariant but exactly the complement of an $\hat{E}_\varphi$-class. This is not specific to $\hat{E}_\varphi$, and more generally we have:

LEMMA 5.4.6. *If E is an almost Σ_1^1 equivalence relation with Π_1^1 domain then E is the restriction of a Σ_1^1 equivalence relation $\hat{E}$ to the complement of an $\hat{E}$-class. In particular E is thin (has countably many classes) iff $\hat{E}$ is thin (has countably many classes)*

PROOF. Let C be the domain of E. Pick any Σ_1^1 equivalence relation $\hat{E}_0$ on a Σ_1^1 set $\hat{C}$ such that C is $\hat{E}_0$-invariant and E is the restriction of $\hat{E}_0$ to C; then define $\hat{E}$ by: $\hat{E} = \hat{E}_0 \cup ((\hat{C} \setminus C) \times (\hat{C} \setminus C))$. □

The converse of the first part of Lemma 5.4.6 is not true: the existence of $\hat{E}$ implies only that E is almost Σ^1_1 but only with $\mathbf{\Pi}^1_1$ (not necessarily Π^1_1) domain.

We shall now prove the following variation of Theorem 5.4.2:

THEOREM 5.4.7. *Let ξ be an ordinal which is countable in L. If $\aleph_\xi^L < \aleph_1$ then any almost Σ^1_1 equivalence relation with Π^1_1 domain and all equivalence classes in $\mathbf{\Sigma}^*_\xi$, satisfies the Perfect Set Theorem.*

PROOF. Observe that for $\xi = 0$ we have $\mathbf{\Sigma}^*_0 = \mathbf{\Sigma}^0_1$, and the result is obvious in this case. The following lemma will prove a parametrized version of the case $\xi = 1$, which we need for the general case; we recall that $\mathbf{\Sigma}^*_1 = \mathbf{\Sigma}^0_3$.

LEMMA 5.4.8. *Let $\alpha \in \omega^\omega$ and assume that $\aleph_1^{L(\alpha)} < \aleph_1$. Let E be an almost $\mathbf{\Sigma}^1_1$ equivalence relation with domain a $\mathbf{\Pi}^1_1$ set with $\mathbf{\Pi}^1_1$ code in $L(\alpha)$, and all equivalence classes in $\mathbf{\Sigma}^0_3$. If E is thin then E has countably many equivalence classes.*

PROOF. Let C be the domain of E.

CLAIM . *C has $\mathbf{\Sigma}^0_3$ approximations.*

Proof: Let A be any $\mathbf{\Sigma}^1_1$ subset of C, and consider the equivalence relation E_A obtained by restricting E to A: Since E is almost $\mathbf{\Sigma}^1_1$ then $E = \hat{E} \cap (C \times C)$ for some $\mathbf{\Sigma}^1_1$ equivalence relation $\hat{E}$, hence $E_A = E \cap (A \times A) = \hat{E} \cap (A \times A)$ is also $\mathbf{\Sigma}^1_1$. Moreover since E is thin then E_A is a fortiori thin too. Finally each E_A-equivalence class is a *relatively* $\mathbf{\Sigma}^0_3$ subset of A, hence by the boldface version of Theorem 5.4.2 for $\xi = 0$, we can conclude that E_A has countably many equivalence classes; so $A = \bigcup_{n\in\omega} E_A(x_n)$ for some sequence $(x_n)_{n\in\omega}$ in A; consequently $A \subset B := \bigcup_{n\in\omega} E(x_n) \subset C$ and it follows from the hypothesis that B is $\mathbf{\Sigma}^0_3$. ◇

Since C has a $\mathbf{\Pi}^1_1$ code in $L(\alpha)$ and $\aleph_1^{L(\alpha)} < \aleph_1$ then by Theorem 5.4.9, C is $\mathbf{\Sigma}^0_3$. Hence the equivalence relation E is $\mathbf{\Sigma}^1_1$, so by a second application of the boldface version of Theorem 5.4.2 for $\xi = 0$, E has countably many equivalence classes. □

We also need the following extension of Theorem 5.1.3 that we admit, and which follows from Theorem 4.4.4 and the arguments of Theorem 5.1.3.

LEMMA 5.4.9. *Let $\alpha \in \omega^\omega$ and ξ be an ordinal which is countable in $L(\alpha)$, and assume that $\aleph_\xi^{L(\alpha)} < \aleph_1$. If C is a $\mathbf{\Pi}^1_1$ set with $\mathbf{\Pi}^1_1$ code in $L(\alpha)$ and has $\mathbf{\Gamma}_\xi$ approximations then C is in $\mathbf{\Gamma}_\xi$.*

The other ingredient of the proof of Theorem 5.4.7 is the general pullback method developed by Stern in [**26**], precisely for the proof of Theorem 5.4.2. For the convenience of the reader we summarize this method in next abstract lemma. This simple but opaque formulation hides in fact the real arguments behind, which are on one hand Silver's Theorem for Borel equivalence relations, and on the other hand an analysis developed by Stern of codes of Borel sets which appear in generic extensions of L.

LEMMA 5.4.10. *(Stern) Let $n \in \omega$ and let ξ be an ordinal which is countable in L. If $\aleph_\xi^L < \aleph_1$ then there exists a Borel isomorphism $f : \omega^\omega \to \omega^\omega$ such that if A is any $\mathbf{\Sigma}^0_{1+\xi+n}$ equivalence class relatively to an arbitrary thin Σ^1_1 equivalence relation, then $f(A)$ is $\mathbf{\Sigma}^0_{1+n}$.*

In fact in [**26**] Stern states all results for Σ^1_1 equivalence relations with *all* equivalence classes in some given Baire class $\mathbf{\Gamma}$, but the reader can easily check that this strong hypothesis is not used, and all arguments can be localized to *any* equivalence class in $\mathbf{\Gamma}$.

Proof of Theorem 5.4.7: We distinguish two cases:

Case 1: *ξ is limit, so $\mathbf{\Sigma}^*_\xi = \mathbf{\Sigma}^0_{1+\xi}$*

Fix a Σ^1_1 equivalence relation $\hat{E}$ as in Definition 5.4.4. Apply Lemma 5.4.10 to the equivalence relation $\hat{E}$, with $n = 0$ and the ordinal ξ, and consider the equivalence relation $E' = (f \times f)(E)$. Since any E-equivalence class is also an $\hat{E}$–equivalence class, then any E'-equivalence class is $\mathbf{\Sigma}^0_1$, and obviously E' has at most countably many classes; and since f is one-to-one then the same holds for E.

Case 2: *$\xi = \eta + 1$ is successor, so $\mathbf{\Sigma}^*_\xi = \mathbf{\Sigma}^0_{1+\xi+1} = \mathbf{\Sigma}^0_{1+\eta+2}$.*

Choose a code $\alpha \in \omega^\omega$ for the countable ordinal $\aleph^L_\eta$ such that $\aleph^L_\xi = \aleph^{L(\alpha)}_1$. Let C be the domain of E, and fix $\hat{E}$ as in Lemma 5.4.6. Apply Lemma 5.4.10 *in the universe* $L(\alpha)$, to $\hat{E}$ with $n = 2$ and the ordinal η, to get a Borel isomorphism f definable in $L(\alpha)$. Now consider (in the universe) the equivalence relation $E' = (f \times f)(\hat{E})$: it is clear that $\hat{E}'$ is almost $\mathbf{\Sigma}^1_1$ and its domain $C' = f(C)$ has a $\mathbf{\Pi}^1_1$ code in $L(\alpha)$ (since f is definable in $L(\alpha)$), finally by Lemma 5.4.10, E'-equivalence classes are all $\mathbf{\Sigma}^0_3$. Since $\aleph^{L(\alpha)}_1 = \aleph^L_\xi < \aleph^L_1$ then by Lemma 5.4.8 E' has at most countably many classes, hence E also. □

REMARKS 5.4.11. a) The equivalence relation constructed by Sami in Theorem 3.5 of [**23**] shows that the previous result is best possible. One can also consider the equivalence relation associated to the Π^1_1-norm φ_ξ of Theorem 5.2.1.

b) In [**25**] and [**26**] Stern obtains similar results in slightly different contexts, but strictly speaking his results are incomparaable with ours. In fact he obtains the same conclusion under weaker descriptive hypotheses and stronger set theoretical assumptions:

– In [**25**] he studies Σ^1_1 equivalence relations with uncountably many (instead of all) $\mathbf{\Pi}^0_2$ equivalence classes. In this context he obtains (Theorem 2.4) the same conclusion assuming that $\aleph^L_1 < \aleph_1$. But his proof (based on a Baire category argument) does not extend even for the case where *all* equivalence classes of some Π^1_1 set are $\mathbf{\Sigma}^0_2$, while under the same assumption our result covers the case where all equivalence classes of some Π^1_1 set are $\mathbf{\Sigma}^0_3$.

– In [**26**] he considers (Theorem 6.1) Σ^1_2 equivalence relations with Δ^1_2 domain; but then to obtain the same conclusion for relations with $\mathbf{\Pi}^0_2$ equivalence classes the argument necessitates that $\aleph^L_4 < \aleph_1$.

CHAPTER 6

Proof of The Main Result

This chapter is entirely devoted to the proof of next result. We recall that $\Gamma_\xi = D(\Sigma^0_{1+\xi+1})$ if ξ is successor, and $\Gamma_\xi = \Sigma^0_{1+\xi+1}$ if ξ is limit.

Main Theorem. *For any recursive ordinal ξ:*

$$\aleph_\xi^L < \aleph_1 \implies \mathit{Lift}(\Pi^1_1, \Gamma_\xi)$$

6.1. Sketch of the proof

6.1.1. General scheme. The proof of Main Theorem is very long and this for several reasons. A first complication comes from the fact that the definition of the classes Γ_ξ is not uniform and depends on whether ξ is successor or limit; and one has to proceed by separating the two cases. In fact the proofs in both cases will follow the same general scheme, but will diverge at several technical but crucial details. More precisely dealing with say a limit ordinal ξ creates difficulties which we will manage to solve because we restrict ourselves to $\Sigma^0_{1+\xi+1}$ sets; and conversely dealing with $D(\Sigma^0_{1+\xi+1})$ sets will create difficulties of some *other type* which we will overcome because ξ is successor.

Also all attempts we tried for a uniform treatment of both cases were unsuccessful and led to extremely heavy and artificial statements. However it happens that the arguments of the two proofs possess a kind of "common part" which is already non trivial, and corresponds to a particular instance of Main Theorem to which we will refer as the "*basic case*". So our plan will be the following:

(1) Prove the basic case:

($\star$) *For ξ successor:* $\aleph_\xi^L < \aleph_1 \implies \mathit{Lift}(\Pi^1_1, \Sigma^0_{1+\xi+1})$

(2) Indicate the needed modifications in the proof of ($\star$) to cover the limit case:

For ξ limit: $\aleph_\xi^L < \aleph_1 \implies \mathit{Lift}(\Pi^1_1, \Sigma^0_{1+\xi+1})$.

(3) Indicate the needed modifications in the proof of ($\star$) to cover the general successor case:

For ξ successor: $\aleph_\xi^L < \aleph_1 \implies \mathit{Lift}(\Pi^1_1, D(\Sigma^0_{1+\xi+1}))$

We again point that the modifications in steps 2) and 3) will be of totally different nature; moreover by Theorem B of the introduction they cannot be realized simultaneously.

Finally let us recall the following weak form of *Lift*(Λ,Γ):

Lift$^{(00)}(\Lambda,\Gamma)$: *"For any $X_0 \subset 2^\omega \times 2^\omega$ in Δ^1_1, any $Y \subset 2^\omega$ in Γ, any $Z \subset 2^\omega$ in Λ, if for any compact subset K of Y there exists $z \in Z$ such that $K \times \{z\} \subset X_0$, then Y has a continuous lifting in $X := X_0 \cap (2^\omega \times Z)$."*

As observed in Remark 4.4.2 a) for any class $\Gamma \subset \Pi^1_1$ we have:

$$\mathit{Lift}^{(00)}(\Pi^1_1,\Gamma) \iff \mathit{Lift}(\Pi^1_1,\Gamma)$$

This simple equivalence is fundamental, and for the proofs of the various instances of Main Theorem discussed above, we shall always work with the form *Lift*$^{(00)}$ which will reveal itself to be much more adapted than *Lift*.

6.1.2. Sketch of the proof of the basic case. Thus assuming $\aleph^L_\xi < \aleph_1$ we want to show that *Lift*$(\Pi^1_1, \Sigma^0_{1+\xi+1})$ holds, but as we explained above we will show that *Lift*$^{(00)}(\Pi^1_1, \Sigma^0_{1+\xi+1})$ holds.

So we fix X_0 a Δ^1_1 subset of $2^\omega \times 2^\omega$, Y a $\Sigma^0_{1+\xi+1}$ subset of 2^ω and Z a Π^1_1 subset of 2^ω, with the property that for any compact subset of Y there exists $z \in Z$ such that $K \times \{z\} \subset X_0$; and we have to prove that Y has a continuous lifting in $X := X_0 \cap (2^\omega \times Z)$.

This will be achieved if we can, as in the proof of Lemma 4.1.4, introduce a game G^* with the following properties:

(0) *G^* is determined.*

(1) *Any winning strategy for Player I in G^* "defines" a compact subset of Y which has no constant lifting in X.*

(2) *Any winning strategy for Player II in G^* "defines" a continuous lifting of Y in X.*

Notice that the game G_0 of Lemma 4.1.4, satisfies already requirements (1) and (2), but since $X = X_0 \cap (2^\omega \times Z)$ is no more Borel one cannot a priori ensure (0).

The natural idea is then to try some modification of this game; but first let us recall the definition of G_0.

a) The game G_0: In the game G_0 *Player I* and *Player II* construct, in a Lipschitz way, finite sequences s_n and t_n in Seq(2), and *Player II* wins an infinite run if

$$y \notin Y \quad \text{or} \quad (y,z) \in X$$

where as usual $y = \bigcup_n s_n$ and $z = \bigcup_n t_n$.

For a better analysis of the situation let us split the condition "$(y,z) \in X$" into two subconditions: (a) "$(y,z) \in X_0$" which is Borel; (b) "$z \in Z$" which is Π^1_1.

A standard way to "transform" such a game into a Borel game, is then to "unfold" the non-Borel condition (b) in a Borel, generally closed, condition. Usually this is achieved by fixing a sequential tree T on $2 \times \omega_1$ such that Z is the projection of $\lceil T \rceil$ on the first factor 2^ω, and by asking *Player II* to choose at each move in addition of $t_n \in$ Seq(2) some $\tau_n \in \mathrm{Seq}(\omega_1)$ with: (i) $|t_n| = |\tau_n|$; (ii) $(t_n, \tau_n) \in T$; (iii) $(t_{n-1}, \tau_{n-1}) \prec (t_n, \tau_n)$; so that in any infinite run, $(z, \bar{\tau}) := \bigcup_n (t_n, \tau_n) \in \lceil T \rceil$ which ensures automatically that $z \in Z$.

Unfortunately this standard procedure cannot work in the present situation, even in the nice case where X_0 is Π^0_1 and Y is the projection of X. More precisely if *Player II* wins the modified game then a winning strategy yields continuous functions $f : Y \to Z$ and $\varphi : Y \to [T]$ such that for every $y \in Y$ $(y, f(y)) \in X$ and $f(y)$ is the projection of $\varphi(y)$. Since $[T]$ is a complete metric space, φ and f extend to continuous $\tilde{\varphi}$ and $\tilde{f}$ both defined on a $\mathbf{\Pi}^0_2$ set $\tilde{Y}$. Then the set

$$\{y \in \tilde{Y} : (y, \tilde{f}(y)) \in X \text{ and } \tilde{f}(y) \text{ is the projection of } \tilde{\varphi}(y)\}$$

is $\mathbf{\Pi}^0_2$, contains Y and is contained in the projection of X, hence is equal to Y. So Y is necessarily $\mathbf{\Pi}^0_2$ (and in this case the problem is already settled by former results !).

So we will introduce a game G with a less trivial unfolding of the condition "$z \in Z$". In particular we shall not ask *Player II* to play the witness $\bar{\tau}$ in a Lipschitz way, but in a more general Borel way. In fact what we shall really unfold is the joint condition: "$y \notin Y$ *or* $z \in Z$", in the sense that the rules of the game G will be such that the witness $\bar{\tau}$ will be constructed only if $y \in Y$.

b) The game G: The definition of this game will in fact depend on a number of parameters that we have to fix first.

Fix a sequential tree T as above. Then identifying, as usual, Y to a $\mathbf{\Sigma}^0_{1+\xi+1}$ subset of $[Ext]$, fix $(\hat{Y}, R, \overrightarrow{R})$ a regular $\mathbf{\Sigma}^0_2$-representation of class ξ of the $\mathbf{\Sigma}^0_{1+\xi+1}$ set Y (see 1.7.4). Finally fix a double-tree representation (R^+, R^-) of $\hat{Y} \subset [R]$ given by Proposition 2.2.2 (with the same configurations of the double chains). Notice that since Z is Π^1_1 then by classical results (see [**18**]) we can suppose that $T \in L$. Again by general absoluteness arguments, since Y is in fact $\Sigma^0_{1+\xi+1}$ one can can also suppose that $(\hat{Y}, R, \overrightarrow{R}, R^+, R^-) \in L$ too so that the game G defined next will also be in L.

Rules: In the game G the players choose at each move finite sequences s_n and t_n in Seq(2) as in G_0; and in addition *Player II* has also to choose some $\tau_n \in \text{Seq}(\omega_1)$ with:

(i) $|\tau_n| \leq |t_n|$

(ii) (t_n, τ_n) *is compatible with* T *(i.e. if* $k = |\tau_n|$ *then* $(t_k, \tau_n) \in T$*)*

(iii) If s_m *is the* R^+*-predecessor of* s_n *then* $\tau_n \succ \tau_m$ *and* $|\tau_n| = |\tau_m| + 1$ *where by convention* $s_{-1} = \tau_{-1} = \emptyset$.

Win condition: *Player II* wins the infinite run $(s_n, t_n, \tau_n)_{n \geq 0}$ in G if

$$y \notin Y \text{ or } (y, z) \in X_0$$

where y and z are as above.

In particular G is a Borel game and so determined. Moreover if $y \in Y$ then y "contains" an infinite R^+-branch $(s_{m_j})_{j \geq 0}$, and it follows then from Rule (iii) that $(t_{m_j}, \tau_{m_j})_{j \geq 0} \in [T]$ which ensures as in Remark a) above that $z \in Z$, hence that $(y, z) \in X$. One can easily derive from these observations that, as in G_0, if *Player II* wins the game G then Y has a continuous lifting in X.

Thus G realizes requirements (0) and (2) above. Unfortunately this is not the case for (1), and to overcome this obstacle we need a new modification of completely other type which will enable us to ensure (1).

c) The game G^*: Informally speaking the game G^* is the "same" as G, except that in G^* we allow *Player II* to announce the τ_n with an arbitrary delay. On the other hand to win an infinite run in G^*, *Player II* will have to announce infinitely many τ_n. To formalize this in a simple way, observe first that if $\tau_n \in \mathrm{Seq}(\omega_1)$ appears in any licit position in G then necessarily $\tau_n \neq \emptyset$. We now define G^* as follows:

Rules: In G^* the players choose at each move s_n and t_n in Seq(2) as in G_0; in addition *Player II* has also to choose some $\tau_n \in \mathrm{Seq}(\omega_1)$ with:

(i) $|\tau_n| \leq |t_n|$

(ii) (t_n, τ_n) *is compatible with* T

(iii*) *If* $(m_i)_{i\leq k}$ *is the increasing enumeration of the set* $\{m \leq n : \tau_m \neq \emptyset\}$ *and* s_j *is the* R^+*-predecessor of* s_k *then* $\tau_{m_k} \succ \tau_{m_j}$ *and* $|\tau_{m_k}| = |\tau_{m_j}| + 1$

Notice that if $\tau_n = \emptyset$ then Rule *(iii*)* is already fulfilled at stage m_k, so $\tau_n = \emptyset$ can be interpreted as a "*pass*".

Win condition: *Player II* wins the infinite run $(s_n, t_n, \tau_n)_{n\geq 0}$ in G^* if

$$y \notin Y \text{ or } \big((y,z) \in X_0 \text{ and } \{n : \tau_n \neq \emptyset\} \textit{ is infinite}\big)$$

Hence G^* is also a Borel game, and again one easily checks that as for G requirements (0) and (2) are fulfilled, and we are reduced to prove (1).

So fix a winning strategy σ for *Player I*: we have to construct a compact subset K of Y which has no constant lifting in X. Notice that in the case of G_0 (see the proof of Lemma 4.1.4) one can take for K the set of all reals y constructed by *Player I* in all infinite runs compatible with σ, and in this case the compactness of K follows directly from the fact that in G_0, at each stage *Player II* has only finitely many possible moves. Of course this is no more the case in G^*, but we shall show that this is still the case "up to some equivalence relation" E which is reducible to the equality relation on a set of cardinality $\aleph_\xi^L$, hence admits at most $\aleph_\xi^L$ equivalence classes. Then using the assumption that $\aleph_\xi^L < \aleph_1$ we shall prove that the set of reals y constructed by *Player I* in sufficiently many (but not all) infinite runs compatible with σ is a compact set K, which will be necessarily contained in Y by the win condition, and which is sufficiently large to ensure requirement (1). The precise definitions of the equivalence relation E and the compact set K are extremely technical, and we shall not try to describe them at this level of generality of the discussion.

6.1.3. Sketch of the proof of the general case. For the limit case, if we repeat the same arguments as above then we finish up with an equivalence relation with $\aleph_{\xi+1}$ equivalence classes, and the assumption "$\aleph_\xi < \aleph_1$" is not sufficient to pursue the argument. However using the fact that in the limit case the resolution family $\overrightarrow{R}$ can be taken *uniform*, one can replace the equivalence relation E by a coarser relation $E^{(\mathrm{u})}$ with precisely $\aleph_\xi$ equivalence classes and which will do the job.

For the general successor case the situation is more complicated, and to cover $\mathit{Lift}(\Pi^1_1, D(\Sigma^0_{1+\xi+1}))$ one has to modify the game G^* itself, and introduce a new game G^{**}. These modifications are related to the technical aspects of the double-tree representation of $D(\mathbf{\Sigma}^0_2)$ sets developped in Section 2.3, namely the combinatorial properties of the double chains in this case.

Also to avoid long repetitions we shall study in next section, a family of games for which we will describe some general constructions that we will apply later on to the games G^* and G^{**} involved in each case.

6.2. Labeling games with delay

In all this section games are considered merely by their *rules* without any reference to a win condition. In other words what we are really studying here are sequential trees satisfying some combinatorial conditions. Still, since all applications will concern games, we shall develop all the material of this section in this context. In fact this study concerns only games of some specific type. For a better intuition, the reader should have in mind the games G and G^* considered in the previous section.

6.2.1. Labeling games: Let Ω be an arbitrary set. A labeling game in Ω (implicitly *"for Player II"*) is a game in which the players proceed as follows:

$$\mathscr{L}(\Omega) : \begin{cases} I: \quad s_0 \qquad\qquad s_1 \qquad\qquad \dots \quad s_n \qquad\qquad \dots \\ II: \qquad (t_0,\tau_0) \qquad (t_1,\tau_1) \quad \dots \qquad (t_n,\tau_n) \quad \dots \\ \\ \quad \text{with:} \quad s_n \ , \ t_n \in 2^{n+1} \ , \ s_n \succ s_{n-1} \ , \ t_n \succ t_{n-1} \ , \ \tau_n \in \Omega \end{cases}$$

and possibly some additional rules. We shall say that τ_n is the label of (s_n, t_n).

We shall identify a position u in G with the triple :

$$(s,t,\theta) \in 2^{m+1} \times 2^{n+1} \times \Omega^{n+1}$$

where: $s = s_m$ is the last move of *Player I* in u; $t = t_n$ is the last Seq(2)-move of *Player II* in u; $\theta = (\tau_m)_{m \leq n}$ is the sequence of all Ω-moves of *Player II* in u. So $m = n-1$ or n, depending whether u is *I*-position or *II* position. We shall set then:

$$\varphi(u) := (s,t) \in 2^{m+1} \times 2^{n+1} \qquad \text{and} \qquad \psi(u) := \theta \in \Omega^{n+1}$$

and identify u with the pair $(\varphi(u), (\psi(u))$. In particular if u is a *II*-position then we can, via obvious identifications, view $\varphi(u) \in 2^{n+1} \times 2^{n+1}$ as an element of $(2\times 2)^{<\omega}$.

DEFINITION 6.2.2. For any tree relation R on Seq(2×2), if u is a *II*-position with $h_R(\varphi(u)) = k$ and

$$\varphi^{(R)}(u) := (s_{m_j}, t_{m_j})_{j \leq k} \in ((2 \times 2)^{<\omega})^{k+1}$$

is the increasing enumeration of the R-branch of $\varphi(u)$, we define:

$$\psi^{(R)}(u) := (\tau_{m_j})_{j \leq k} \in \Omega^{k+1}$$

the *labeling of the R-branch* of $\varphi(u)$.

We shall say that G is an *R-labeling game in Ω* if there exists a sequential tree $\mathscr{T}$ on Seq$((2 \times 2)) \times \Omega$ such that for all $u \in (\text{Seq}((2 \times 2)) \times \Omega)^{<\omega}$:

$$u \text{ is a legal II-position in } G \iff (\varphi^{(R)}(u), \psi^{(R)}(u)) \in \mathscr{T}$$

We shall then say that G is an R-labeling game in Ω, with rule $\mathscr{T}$.

So more informally, an R-labeling game is a labeling game in which the rules *for Player II* depend only on the R-branch $\varphi^{(R)}(u)$ of $\varphi(u)$ and its labeling $\psi^{(R)}(u)$. Notice that $\varphi^{(R)}(u)$ is entirely determined by its last element $\varphi(u)$; but this is not

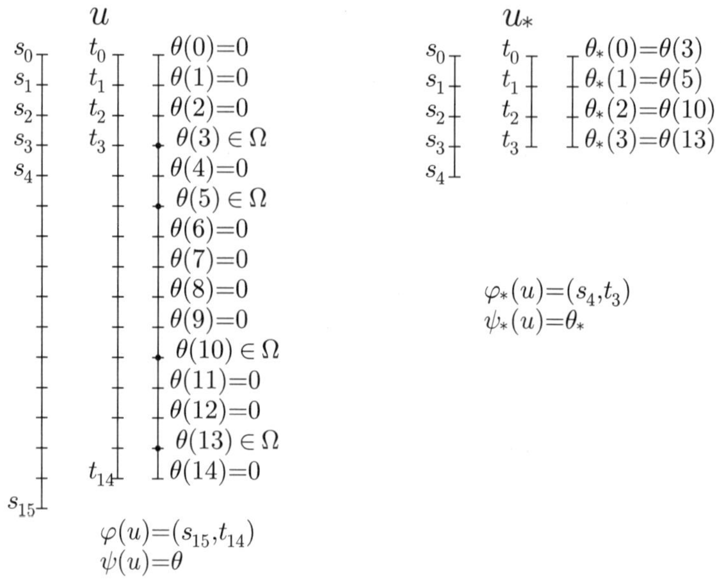

FIGURE 6.1

the case for $\psi^{(R)}(u)$. Obviously the game G considered in Remark 6.1.2 b) is an R^+-labeling game in $\Omega = \mathrm{Seq}(\omega_1) \setminus \{\emptyset\}$; but for the proof of Theorem 6.3.1 we will have to consider an R-labeling game where R is the tree generated by the double-tree (R^+, R^-).

6.2.3. Labeling games with delay: Let Ω be an arbitrary set and 0 be a fixed "element" *not in* Ω, that we shall call the <u>external element</u>, and set: $\Omega^* = \Omega + \{0\}$.

For any $\theta \in (\Omega^*)^{<\omega}$ we define the <u>support</u> of θ:

$$\mathrm{supp}(\theta) = \{m \in \mathrm{dom}(\theta) : \ \theta(m) \neq 0\}$$

and denote by $\theta_* \in \mathrm{Seq}(\Omega)$ the sequence obtained by deleting from θ all elements with value 0 (*i.e.* $\theta_* = \theta \circ \imath$ where $\imath$ is the increasing enumeration of $\mathrm{supp}(\theta)$).

DEFINITION 6.2.4. To any position $u = (s, t, \theta) \in 2^{m+1} \times 2^{n+1} \times \Omega^{n+1}$ in a labeling game in Ω^* (*i.e.* satisfying $\mathscr{L}(\Omega^*)$), we can canonically associate

$$u_* = (s_{|q}, t_{|p}, \theta_*) \in 2^q \times 2^p \times \Omega^p$$

where θ_* is as above, $p = |\theta_*|$ and $q = p+1$ or p, depending whether u is *I*-position or *II*-position. We then set:

$$\varphi_*(u) := (s_q, t_p) \quad \text{ and } \quad \psi_*(u) := \theta_*$$

and again if u is a *II*-position then we shall view $\varphi_*(u)$ as an element of $(2 \times 2)^{<\omega}$. Obviously u_* satisfies $\mathscr{L}(\Omega)$ hence can be interpreted as a position in some labeling game in Ω.

Now given any tree relation R on $\mathrm{Seq}(2 \times 2)$, we define for any *II*-position:

$$\varphi_*^{(R)}(u) := \varphi^{(R)}(u_*) \quad \text{ and } \quad \psi_*^{(R)}(u) := \psi^{(R)}(u_*)$$

Notice that if $u \prec v$ then $u_* \prec v_*$; in particular if u is a *I* (*II*)-position then u_* is a *I* (*II*)-position.

Hence given any labeling game G in Ω, we can define a labeling game G^* in Ω^* as follows. For any position u satisfying $\mathscr{L}(\Omega^*)$:

$$(\varphi(u), \psi(u)) \text{ is a legal position in } G^* \iff (\varphi_*(u), \psi_*(u)) \text{ is a legal position in } G$$

We shall say that G^* is the game obtained from G by *label delay*; if moreover G is an R-labeling game in Ω we shall then say that G^* is an R-labeling game in Ω^*.

It is clear that the game G^* considered in 6.1.2 c) is of this form with $\Omega = \mathrm{Seq}(\omega_1) \setminus \{\emptyset\}$ and $0 = \emptyset$ (so $\Omega^* = \mathrm{Seq}(\omega_1)$); more precisely G^* is obtained by label delay from the game G considered in 6.1.2 b).

Finally as observed in the particular case 6.1.2 c), one should interprete 0 in a general labeling game with delay G^*, as a *"pass"* for the labeling operation, so we can more intuitively imagine that in a position u in G^*, *Player II* is in fact labeling only an initial segment of $\varphi(u)$, and $\varphi_*(u)$ is precisely the largest element $(s,t) \preceq \varphi(u)$ labeled in u.

6.2.5. General notations. *From now on by a "position" we shall mean a "II-position".*

We fix:

$$\begin{cases} - (\Omega, <) \text{ a well-ordered set, with an external element } 0 \\ - G^* \text{ an } R\text{-labeling game with delay in } \Omega \\ - \overrightarrow{R} = (R^{(\eta)})_{\eta \leq \xi} \text{ a resolution family for } R \text{ in Ext} \end{cases}$$

So:

$$R^{(0)} = Ext\ ,\ R^{(\eta+1)} \subset R^{(\eta)}\ ,\ R^{(\lambda)} = \bigcap_{\zeta < \lambda} R^{(\zeta)} \text{ if } \lambda \text{ is limit}\ ,\ R^{(\xi)} = R$$

For the main result of this section we shall also require that the resolution family $\overrightarrow{R}$ is distinguished, but for the moment we only need the monotonicity and continuity of $\overrightarrow{R}$.

Let W denote the set of all positions in G^*. Our aim is to give a transfinite analysis of W, induced by the resolution family $\overrightarrow{R}$.

Following the conventions of sections 6.2.1 and 6.2.3 above, we view W as a subset of $(2 \times 2 \times \Omega^*)^{<\omega}$. Notice that W is then identified to a sequential tree on $2 \times 2 \times \Omega^*$, and we shall denote as usual by $\prec$ and $|.|$ the extension relation (which coincides with the extension relation in the game) and the length mapping, inherited from $(2 \times 2 \times \Omega^*)^{<\omega}$.

For all $\eta \leq \xi$ we set:

$$\psi_\eta(u) := \psi_*^{(R^{(\eta)})}(u) \in \mathrm{Seq}(\Omega)$$

which we recall is the sequence defined by the labeling in Ω (and not in Ω^*) of the $R^{(\eta)}$-branch of $\varphi_*(u)$; in particular $\psi_0(u) = \psi_*(u) = \psi(u)_*$. We shall also consider the sets:

$$J_\eta(u) := \{m \leq |u_*| : \quad \varphi_*(u_{|m})\ R^{(\eta)}\ \varphi_*(u)\}$$

In particular:

$$\forall m \in J_\eta(u), \quad \psi_\eta(u_{|m}) \preceq \psi_\eta(u)$$

Notice that if λ is limit then $J_\lambda(u) = \bigcap_{\zeta<\lambda} \downarrow J_\zeta(u)$, and since all these sets are finite then:

LEMMA 6.2.6. *If λ is limit then for all $u \in W$ there exists $\eta < \lambda$ such that $J_\lambda(u) = J_\eta(u)$, and therefore $\psi_\eta(u) = \psi_\lambda(u)$.*

We recall that by assumption $(\Omega, <)$ is a well-ordered set, hence the set $\mathrm{Seq}(\Omega)$ is canonically well-ordered by the lexicographical ordering induced by $<$ (see [**4**], 2.4) that we also denote by $<$. In fact in all applications Ω will be already of the form $\mathrm{Seq}(\kappa)$ for some ordinal κ, so that in this case $\mathrm{Seq}(\Omega) = \mathrm{Seq}(\mathrm{Seq}(\kappa))$ will be endowed with the "double" lexicographical wellordering induced by κ.

DEFINITION 6.2.7. We define inductively three families: $(V_\eta)_{\eta\leq\xi}$ of subsets of W, $(\tilde{E}_\eta)_{\eta\leq\xi}$ of equivalence relations on W, and $(E_\eta)_{\eta\leq\xi}$ of equivalence relations on $\mathrm{Seq}(\Omega)$ by:

– *For* $\eta = 0$

(i) $V_0 = W$

(ii) $\theta \, \tilde{E}_0 \, \theta' \iff |\theta| = |\theta'|$ and $\{(i,j) : \theta(i) = \theta(j)\} = \{(i,j) : \theta'(i) = \theta'(j)\}$

(iii) $u \, \tilde{E}_0 \, v \iff \varphi(u) = \varphi(v)$, $\mathrm{supp}(\psi(u)) = \mathrm{supp}(\psi(v))$ and $\psi_0(u) = \psi_0(v)$

– *For* $\eta < \xi$:

(iv) $V_{\eta+1} = \big\{u \in V_\eta : \forall m \in J_{\eta+1}(u),$

$$\psi_\eta(u_{|m}) = \min\{\psi_\eta(u') : u' \in V_\eta \, , \; u' \, \tilde{E}_\eta \, u_{|m} \, , \; \psi_{\eta+1}(u') = \psi_{\eta+1}(u_{|m})\} \big\}$$

(v) $\theta \, E_{\eta+1} \, \theta' \iff \theta \, E_\eta \, \theta'$ and $\forall k \quad \theta_{|k} \, F_{\eta+1} \, \theta'_{|k}$ and $\theta'_{|k} \, F_{\eta+1} \, \theta_{|k}$

(vi) $u \, \tilde{E}_{\eta+1} \, v \iff u \, \tilde{E}_\eta \, v$ and $\psi_{\eta+1}(u) \, E_{\eta+1} \, \psi_{\eta+1}(v)$

where $\theta \, F_{\eta+1} \, \theta'$ is defined by :

$$\forall u \in V_{\eta+1} \quad \Big(\theta \preceq \psi_{\eta+1}(u) \Rightarrow \exists u' \in V_{\eta+1} : \theta' \preceq \psi_{\eta+1}(u') \text{ and } u \, \tilde{E}_\eta \, u'\Big)$$

– *For limit* $\lambda \leq \xi$:

(vii) $V_\lambda = \bigcap_{\zeta<\lambda} V_\zeta$

(viii) $\theta \, E_\lambda \, \theta' \iff \forall \zeta < \lambda \quad \theta \, E_\zeta \, \theta'$

(ix) $u \, \tilde{E}_\lambda \, v \iff \forall \zeta < \lambda \quad u \, \tilde{E}_\zeta \, v$

REMARKS 6.2.8. a) The family $(V_\eta)_{\eta\leq\xi}$ is non-increasing.

b) The basic property of $V_{\eta+1}$ is that the labeling mapping $\psi_{\eta+1}$ of the $R^{(\eta+1)}$-branches, is one-to-one on each $\tilde{E}_\eta$-equivalence class. The specific choice of the lexicographical well-ordering in this definition is essential, because of the following compatibility property:

$$\theta < \theta' \implies \theta_{|k} < \theta'_{|k}$$

which means more concretely that we do not change our preference after extension.

c) The mappings $J_\eta : W \to \mathscr{P}(\omega)$ play a fundamental role in this definition. First observe that the mappings φ and φ_* are $\tilde{E}_0$-invariant on W. It follows then that the mappings J_η are also $\tilde{E}_0$ (hence $\tilde{E}_\eta$)-invariant. For a better intuition one should think "$m \in J_\eta(u)$" as "m is an η-good level of restriction for u".

For example even though V_η is not a sequential tree (for the extension relation on positions), it is closed under J_η-restrictions:

$$u \in V_\eta \text{ and } m \in J_\eta(u) \implies u_{|m} \in V_\eta$$

Similarly $\tilde{E}_\eta$ is compatible with J_η-restrictions:

$$u \,\tilde{E}_\eta\, v \text{ and } m \in J_\eta(u) \implies u_{|m} \,\tilde{E}_\eta\, v_{|m}$$

d) The equivalence relations $(\tilde{E}_\eta)_{\eta \leq \xi}$ and $(E_\eta)_{\eta \leq \xi}$ are obviously intimately related. Very roughly speaking two positions u and u' are $\tilde{E}_\eta$-equivalent iff $\psi_\eta(u)$ and $\psi_\eta(u')$ (the labelings of the $R^{(\eta)}$-branches of $\varphi_*(u)$ and $\varphi_*(u')$) are E_η-equivalent. Conversely two elements θ and θ' in $\mathrm{Seq}(\Omega)$ are E_η-equivalent iff *many* pairs of positions v and v' (computed from θ and θ') in which θ and θ' appear as the *beginnings* of $\psi_\eta(v)$ and $\psi_\eta(v')$, are pairwise $\tilde{E}_\eta$-equivalent.

In particular the $\tilde{E}_\eta$-equivalence of two positions u and u' is in fact telling things about many other positions *extending* u and u', or even extending J_η-restrictions of u and u'. This is one of the fundamental properties of these equivalence relations, which is obviously amplified by the inductive aspect of the definitions. For example $u \,\tilde{E}_2\, u'$ means that many pairs (v, v') of positions extending u and u' (or even extending two J_2-restrictions of u and u') are $\tilde{E}_1$-equivalent, which means that many pairs (w, w') of positions extending v and v' (or even extending two J_1-restrictions of v and v') are $\tilde{E}_0$-equivalent, which means that except for the actual values of their labelings w and w' are totally *indiscernible*.

Obviously the equivalence relations E_0 and $\tilde{E}_0$ have countably many classes. In the general case we have the following estimate:

PROPOSITION 6.2.9. *If all parameters $(\Omega, <)$, $\mathscr{T}$, $\overrightarrow{R}$ are in L then each of the equivalence relations $E_{1+\eta}$ and $\tilde{E}_{1+\eta}$ has at most $\aleph^L_{\eta+1}$ classes.*

PROOF. We shall define for all $\eta \leq \xi$, sets Λ_η and $\tilde{\Lambda}_\eta$, and mappings $\Phi_\eta : \mathrm{Seq}(\Omega) \to \Lambda_\eta$ and $\tilde{\Phi}_\eta : W \to \tilde{\Lambda}_\eta$ such that:

$$(1) \qquad \theta \, E_\eta \, \theta' \iff \Phi(\theta) = \Phi(\theta') \quad \text{and} \quad u \,\tilde{E}_\eta\, u' \iff \tilde{\Phi}(u) = \tilde{\Phi}(u')$$

and then show that:

$$(2) \qquad \max\left(\mathrm{card}(\Lambda_{1+\eta}), \mathrm{card}(\tilde{\Lambda}_{1+\eta})\right) \leq \aleph^L_{\eta+1}$$

from which the conclusion of the proposition follows.

We define the families $(\Phi_\eta)_{\eta \leq \xi}$ and $(\tilde{\Phi}_\eta)_{\eta \leq \xi}$ by induction:

– $\Phi_0(\theta) = \{(i,j) \in |\theta| \times |\theta| : \theta(i) = \theta(j)\}$

$\tilde{\Phi}_0(u) = \langle \varphi(u), \mathrm{supp}(\psi(u)), \Phi_0(\psi_0(u)) \rangle$

– For $\eta < \xi$:

$\Phi_{\eta+1}(\theta) = \langle \Phi_\eta(\theta) , \{(k, \tilde{\Phi}_\eta(u)) ; \ u \in V_{\eta+1} \text{ and } \theta_{|k} \preceq \psi_{\eta+1}(u)\} \rangle$

$\tilde{\Phi}_{\eta+1}(u) = \langle \tilde{\Phi}_\eta(u) , \Phi_{\eta+1}(\psi_{\eta+1}(u)) \rangle$

– For limit $\lambda \leq \xi$:

$\Phi_\lambda(\theta) = \{(\zeta, \Phi_\zeta(\theta)) : \zeta < \lambda\}$ and $\tilde{\Phi}_\lambda(u) = \langle \{(\zeta, \tilde{\Phi}_\zeta(u)) : \zeta < \lambda\}, \Phi_\lambda(\psi_\lambda(u)) \rangle$

The verification of (1) is immediate by inspecting the definitions. To prove (2) notice that since $\mathrm{Seq}(\Omega)$ and W are contained in L, and $(\Phi_\eta)_{\eta\leq\xi}$ and $(\tilde{\Phi}_\eta)_{\eta\leq\xi}$ are definable in L, it is enough, by absoluteness, to prove:

$$\text{(3)} \qquad L \models \quad \forall \eta \leq \xi,\ \max(\mathrm{card}(\Lambda_{1+\eta}), \mathrm{card}(\tilde{\Lambda}_{1+\eta})) \leq \aleph_{\eta+1}$$

which using *GCH* in L follows by straightforward induction from next observations:

– $\Lambda_0 \subset \bigcup_{n<\omega} \mathscr{P}(n \times n)$ and $\tilde{\Lambda}_0 \subset \Lambda_0 \times \Lambda_0 \times \Lambda_0$

– $\Lambda_{\eta+1} \subset \Lambda_\eta \times \mathscr{P}(\tilde{\Lambda}_\eta)$ and $\tilde{\Lambda}_{\eta+1} \subset \tilde{\Lambda}_\eta \times \Lambda_{\eta+1}$

– $\Lambda_\lambda \subset \mathscr{P}(\bigcup_{\zeta<\lambda}\{\zeta\} \times \Lambda_\zeta)$ and $\tilde{\Lambda}_\lambda \subset \mathscr{P}(\bigcup_{\zeta<\lambda}\{\zeta\} \times \tilde{\Lambda}_\zeta) \times \Lambda_\lambda$, for λ limit. □

PROPOSITION 6.2.10. *a) If $\theta\, E_\eta\, \theta'$ and $k \leq |\theta|$ then $\theta_{|k}\, E_\eta\, \theta'_{|k}$.*

b) For $\zeta < \eta$, if $u \in V_\zeta$ and $J_\zeta(u) = J_\eta(u)$ then $u \in V_\eta$.

c) If $u\, \tilde{E}_\eta\, v$ and $m \in J_\eta(u)$ then $u_{|m}\, \tilde{E}_\eta\, v_{|m}$.

d) If $u \in V_\eta$ and $m \in J_\eta(u)$ then $u_{|m} \in V_\eta$.

These closure properties are elementary from the Definition 6.2.7, and we will only sketch their proof.

a) : It follows from *(v)* and *(viii)*.

b) : Notice that if $m \in J_\eta(u)$ then $J_\eta(u_{|m}) \subset J_\eta(u)$. Since $J_\beta(u) = J_\eta(u)$ for $\zeta \leq \beta < \eta$, we have $\psi_\beta(u) = \psi_\eta(u)$, hence $u \in V_\beta \Rightarrow u \in V_{\beta+1}$ for $\beta \in [\zeta, \eta[$. Then it follows from *(vii)* that $u \in V_\eta$ whenever $u \in V_\zeta$.

c) : For $m \in J_\eta(u)$ there is a k such that $\psi_\eta(u_{|m}) = \psi(u)_{|k}$. So this follows from *(v)*, *(vi)* and *(ix)*

d) : It follows from *(iv)* and *(ix)* because of the quantifier "$\forall m \in J_{\eta+1}$" in *(iv)*.

The next result states a more fundamental property.

PROPOSITION 6.2.11. *a) If $u, v \in V_\eta$ with $u\, \tilde{E}_\eta\, v$ and $\psi_\eta(u) = \psi_\eta(v)$, then $u = v$.*

b) If $u, v \in V_{\eta+1}$ with $u\, \tilde{E}_\eta\, v$ and $\psi_{\eta+1}(u) = \psi_{\eta+1}(v)$, then $u = v$.

PROOF. We prove part *a)* by induction: For $\eta = 0$, notice that if $u\, \tilde{E}_0\, v$ then $\varphi(u) = \varphi(v)$ and $\mathrm{supp}(\psi(u)) = \mathrm{supp}(\psi(v))$. By definition we have $\psi_0(u) = \psi_*(u)$ and $\psi_0(v) = \psi_*(v)$. If moreover $\psi_0(u) = \psi_0(v)$ then $\psi(u) = \psi(v)$; hence $u = v$.

Suppose that the result is true for η. Assume $u, v \in V_{\eta+1}$ with $u\, \tilde{E}_{\eta+1}\, v$ and $\psi_{\eta+1}(u) = \psi_{\eta+1}(v)$, in particular $u\, \tilde{E}_\eta\, v$ and $\psi_{\eta+1}(u) = \psi_{\eta+1}(v)$. Then it follows from the definition of $V_{\eta+1}$ that $\psi_\eta(u) = \psi_\eta(v)$, and the conclusion follows from the induction hypothesis.

Let λ be a limit ordinal and suppose that the result is true for all $\eta < \lambda$. If $u, v \in V_\lambda$ with $u\, \tilde{E}_\lambda\, v$ and $\psi_\lambda(u) = \psi_\lambda(v)$ then by Lemma 6.2.6 $\psi_\eta(u) = \psi_\eta(v)$ for some $\eta < \lambda$; and since $u, v \in V_\eta$ with $u\, \tilde{E}_\eta\, v$ the conclusion follows from the induction hypothesis.

Part *b)* follows from the argument of *a)* in the successor case. □

Up to now we only proved regularity properties for the families of positions V_η. Our next goal is to show that these families are rich enough. To formulate this property in a precise language we need to introduce two more notions.

6.2.12. Characteristic sequence. The notion that we shall introduce here for the game G^* derives, in fact, from a more general notion for transfinite families of tree relations, that we shall introduce first.

Abstract setting: Let $\overrightarrow{R} = (R^{(\eta)})_{\eta\leq\xi}$ be any resolution family of tree relations on a set A, so satisfying:

$$R^{(\eta+1)} \subset R^{(\eta)} \qquad \text{and} \qquad R^{(\lambda)} = \bigcap_{\zeta<\lambda} R^{(\zeta)} \text{ if } \lambda \text{ is limit}$$

Fix an element $a \in A$; and denote for all $\eta \leq \xi$, by A_η the $R^{(\eta)}$-branch of a. So $(A_\eta)_{\eta\leq\xi}$ is a monotone family of finite sets. The *characteristic sequence* of a is the sequence $\chi(a) := (\xi_i)_{0\leq i\leq n}$ defined by:

$$\begin{cases} \xi_0 = \xi \\ \xi_i = \max\{\zeta \leq \xi_{i-1} : A_\zeta \neq A_{\xi_{i-1}}\} \qquad \text{for } i > 0 \end{cases}$$

Notice that if λ is limit, then $A_\lambda = \bigcap_{\zeta<\lambda} \downarrow A_\zeta = A_{\eta+1}$ for some $\eta < \lambda$. This shows that for $i > 0$, ξ_i is well defined by the last formula, and that except possibly the first element ξ_0, all the other ordinals of the characteristic sequence are *successor* ordinals.

Obviously the sequence $(\xi_i)_{0\leq i\leq n}$ is decreasing, and for all $\eta \in]\xi_{i+1}, \xi_i]$, $A_\eta = A_{\xi_i}$; in particular: $A_{\xi_{i-1}+1} = A_{\xi_i}$

Games setting: For any position u in G^* we define:

$$\bar{\chi}(u) = \chi(\varphi_*(u))$$

that we also call the *characteristic sequence* of u.

We recall that $\varphi_*(u)$ is the largest $(s,t) \preceq \varphi(u)$ which is labeled in u. In particular $\bar{\chi}(u)$ is $\tilde{E}_0$-invariant, and depends only $\varphi_*(u)$, which might be much shorter than $\varphi(u)$; this detail will be crucial for the sequel. Notice also that:

$$\forall\, \eta \in]\xi_{i+1}\,, \xi_i] \quad J_\eta(u) = J_{\xi_i}(u) \;\text{ hence }\; \psi_\eta(u) = \psi_{\xi_i}(u)$$

Moreover we have the following basic property:

$$u\; \tilde{E}_\xi\; v \iff \begin{cases} u\; \tilde{E}_0\; v & \text{with } \bar{\chi}(u)=\bar{\chi}(v) = (\xi_i)_{0\leq i\leq n} \\ \text{and} & \\ \psi_{\xi_i}(u)\; E_{\xi_i}\; \psi_{\xi_i}(v) & \text{for all } i \in [0,n] \end{cases}$$

6.2.13. Potential label. Let G^* be an R-labeling game with delay in Ω, with rule $\mathscr{T}$. We recall that $\mathscr{T}$ is a sequential tree on $2\times 2\times \Omega$, and a position u satisfying $\mathscr{L}(\Omega^*)$ is legal in G^* iff:

$$(\varphi_*^{(R)}(u), \psi_*^{(R)}(u)) \in \mathscr{T}$$

Hence if v is an extension of u in G^* such that $\varphi_*(u)$ is the R-predecessor of $\varphi_*(v)$, then for some $\tau \in \Omega$, with obvious identifications:

$$(\varphi_*^{(R)}(v), \psi_*^{(R)}(v)) = (\varphi_*^{(R)}(u), \psi_*^{(R)}(u))^\frown(\varphi_*(v), \tau) \in \mathscr{T}$$

We shall say that τ is a potential label of $(s,t) \in \mathrm{Seq}(2)$ beyond u if $\varphi_*(u)$ is the R-predecessor of (s,t) and:

$$(\varphi_*^{(R)}(u), \psi_*^{(R)}(u))^\frown((s,t),\tau) \in \mathscr{T}$$

THEOREM 6.2.14. *Assume that the resolution family $(R^{(\eta)})_{\eta\leq\xi}$ is distinguished.*

Let $(s,t) \in \mathrm{Seq}(2)$ with $\chi(s,t) = (\xi_i)_{0\leq i\leq p}$. Suppose that $(w_i)_{0\leq i\leq p}$ is a sequence in V_ξ satisfying for $m_i = |w_i|$:

1) $\varphi_*(w_i)$ is the $R^{(\xi_i)}$-predecessor of (s,t)

2) if $i<j$ then $m_i < m_j$ and $w_j{}_{|m_i}\ \tilde{E}_{\xi_j}\ w_i$

Then for any potential label τ of (s,t) beyond w_0 there exists $v \in V_\xi$ such that:

$$w_0 \prec v \ ,\ \varphi_*(v) = (s,t)\ ,\ \psi_\xi(v) = \psi_\xi(w_0)^\frown\langle\tau\rangle \text{ and } v_{|m_i}\ \tilde{E}_{\xi_i}\ w_i \text{ for all } i > 0$$

PROOF. The proof of Theorem 6.2.14 will make use of the following two general properties that we prove first.

LEMMA 6.2.15. *Let $u,\ u' \in V_{\eta+1}$ be such that $u\ \tilde{E}_{\eta+1}\ u'$. Then for any $v \in V_{\eta+1}$ extending u with $u = v_{|m}$ for some $m \in J_{\eta+1}(v)$, there exists $v' \in V_{\eta+1}$ extending v and such that $v\ \tilde{E}_\eta\ v'$.*

PROOF. Set $\theta = \psi_{\eta+1}(u)$ and $\theta' = \psi_{\eta+1}(u')$; then by assumption $\theta\ E_{\eta+1}\ \theta'$, in particular $\theta\ F_{\eta+1}\ \theta'$; and since $\theta \preceq \psi_{\eta+1}(v)$ then by definition of $F_{\eta+1}$, there exists $v' \in V_{\eta+1}$ such that $\theta' \preceq \psi_{\eta+1}(v')$ and $v\ \tilde{E}_\eta\ v'$; and we shall now show that v' extends u'.

First notice that since $v\ \tilde{E}_0\ v'$ then $m \in J_{\eta+1}(v) = J_{\eta+1}(v') \subset J_\eta(v')$ and $\psi_{\eta+1}(v'{}_{|m}) = \theta'$. So by Proposition 6.2.10 c) and d), $v'{}_{|m} \in V_{\eta+1}$ and $v'{}_{|m}\ \tilde{E}_\eta\ v_{|m} = u\ \tilde{E}_{\eta+1}\ u'$, hence $v'{}_{|m}\ \tilde{E}_\eta\ u'$. Moreover since $\psi_{\eta+1}(v'{}_{|m}) = \psi_{\eta+1}(u')$, it follows then from Proposition 6.2.11 b) that $v'{}_{|m} = u'$. □

LEMMA 6.2.16. *Let $u \in V_\eta$ and let m be such that $\varphi_*(u_{|m})$ is the $R^{(\eta+1)}$-predecessor of $\varphi_*(u)$. If $u_{|m} \in V_{\eta+1}$ then there exists a unique position $v \in V_{\eta+1}$ such that: $\psi_{\eta+1}(v) = \psi_{\eta+1}(u)$, $v\ \tilde{E}_\eta\ u$, and $v_{|m} = u_{|m}$.*

PROOF. The uniqueness follows from Proposition 6.2.11 *b)*. For all $w \in V_\eta$ set:

$$V_\eta(w) = \{w' \in V_\eta :\ w'\ \tilde{E}_\eta\ w\ ,\ \psi_{\eta+1}(w') = \psi_{\eta+1}(w)\}$$

and let $v \in V_\eta(u)$ be such that $\psi_\eta(v)$ is minimal.

Notice that since $\varphi_*(u_{|m})\, R^{(\eta+1)}\, \varphi_*(u)$ then by definition, $m \in J_{\eta+1}(u) \subset J_\eta(u)$, $\psi_{\eta+1}(u_{|m}) = \psi_\eta(u)_{|k}$ for some k; and since $v\ \tilde{E}_0\ u$ the same holds for v. Hence by Proposition 6.2.10 c) and d), $v_{|m} \in V_\eta$, $v_{|m}\ \tilde{E}_\eta\ u_{|m}$ and $\psi_{\eta+1}(u_{|m}) = \psi_{\eta+1}(v_{|m})$. Hence $v_{|m} \in V_\eta(u_{|m})$ and since by assumption $u_{|m} \in V_{\eta+1}$ then by minimality $\psi_\eta(u_{|m}) \leq \psi_\eta(v_{|m})$.

On the other hand by the minimal choice of v in $V_\eta(u)$ we have $\psi_\eta(v) \leq \psi_\eta(u)$, hence by lexicographical ordering we also have $\psi_\eta(v_{|m}) \leq \psi_\eta(u_{|m})$. This proves that $\psi_\eta(v_{|m}) = \psi_\eta(u_{|m})$, and it follows that $v_{|m} \in V_{\eta+1}$; hence by Proposition 6.2.11, $v_{|m} = u_{|m}$. □

Proof of Theorem 6.2.14:

Fix (s,t), $(w_i)_{0\leq i\leq p}$ and τ. Notice first that by hypothesis $\varphi_*(w_0)$ is the R-predecessor of (s,t).

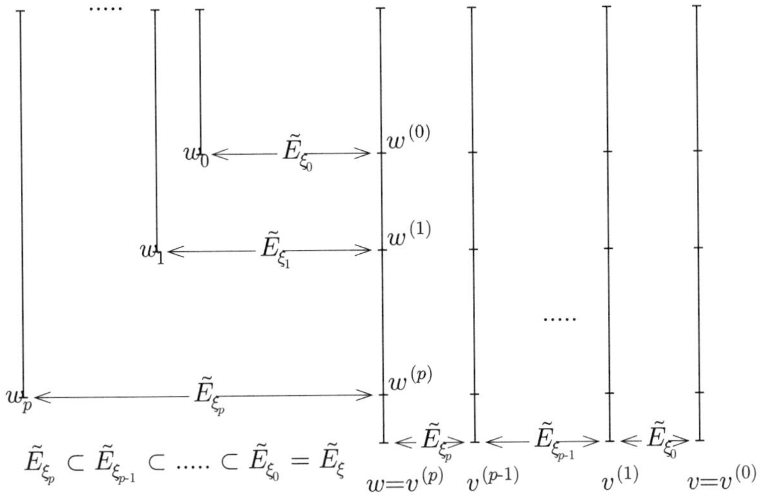

FIGURE 6.2

CLAIM 1. *There exists a sequence* $(w^{(i)})_{0\leq i\leq p}$ *of positions, with* $w^{(0)} = w_0$ *and satisfying for all* $i > 0$ *:*

$$w^{(i)} \in V_{\xi_i} \quad , \quad w^{(i)}\ \tilde{E}_{\xi_i}\ w_i \quad , \quad \textit{and} \quad w^{(i-1)} \prec w^{(i)}$$

Proof: By induction: suppose that $w^{(i-1)}$ is constructed, and notice that for all $i > 0$:

$$\xi_i < \xi_i + 1 \leq \xi_{i-1}$$

Since $w_i \in V_{\xi_i}$ and $\xi_i < \xi$ then $w_i \in V_{\xi_i+1}$; on the other hand by the induction hypothesis, $w^{(i-1)} \in V_{\xi_{i-1}}$ hence $w^{(i-1)} \in V_{\xi_i+1}$. Also by the induction hypothesis, we have $w_{i\,|m_{i-1}}\ \ \tilde{E}_{\xi_{i-1}}\ \ w_{i-1}\ \ \tilde{E}_{\xi_{i-1}}\ \ w^{(i-1)}$ hence a fortiori $w_{i\,|m_{i-1}}\ \ \tilde{E}_{\xi_i+1}\ \ w^{(i-1)}$.

Moreover by assumption we have $\varphi_*(w_i)\ R^{(\xi_i)}\ (s,t)$ and $\varphi_*(w_{i-1})\ R^{(\xi_{i-1})}\ (s,t)$, hence by the definition of χ we also have $\varphi_*(w_{i-1})\ R^{(\xi_i+1)}\ (s,t)$. It follows then by distinction that $\varphi_*(w_{i-1})\ R^{(\xi_i+1)}\ \varphi_*(w_i)$ and by $\tilde{E}_0$ equivalence, $\varphi_*(w_{i\,|m_{i-1}})\ R^{(\xi_i+1)}\ \varphi_*(w_i)$.

So $w_{i\,|m_{i-1}}\ \ \tilde{E}_{\xi_i+1}\ \ w^{(i-1)}$ with $m_{i-1} \in J_{\xi_i+1}(w_i)$; and applying Lemma 6.2.15 we get $w^{(i)} \in V_{\xi_i}$ such that $w^{(i)} \succ w^{(i-1)}$ and $w^{(i)}\ \tilde{E}_{\xi_i}\ u_{m_i}$. ◇

Let $w := w^{(p)}\frown(s,t,\tau)$. By assumption w is a legal position in G^*. Moreover for all i: $w_{|m_i} = w^{(i)} \in V_{\xi_i}$ and $\varphi^*(w_{|m_i})$ is the $R^{(\xi_i)}$-predecessor of $\varphi^*(w) = (s,t)$. But a priori $w \in V_0$, and our aim now is to construct a new position $v \in V_\xi$ with the same properties as w.

CLAIM 2. *There exists a sequence* $(v^{(i)})_{0\leq i\leq p}$ *of positions with* $v^{(p)} = w$ *and satisfying for* $i < p$*:*

$$v^{(i)} \in V_{\xi_i}\ ,\ v^{(i)}{}_{|m_i} = w_{|m_i}\ ,\ \psi_{\xi_i}(v^{(i)}) = \psi_{\xi_i}(w)\ ,\ \ \textit{and}\ \ v^{(i)}\ \tilde{E}_{\xi_{i+1}}\ v^{(i+1)}$$

Proof: By downward induction: for the initial value $i = p$, notice that since $J_0(w) = J_{\xi_p}(w)$ then by Proposition 6.2.10 b), $v^{(p)} \in V_{\xi_p}$; the other clauses are obvious.

Assume that $v^{(i+1)}$ is constructed. Set: $u = v^{(i+1)}$, $\eta = \xi_{i+1}$, $m = m_i$, so that: $u \in V_\eta$ and $m \in J_{\eta+1}(u) = J_{\xi_i}(u)$; then applying Lemma 6.2.16 we get a position $v^{(i)} \in V_{\eta+1}$ such that $v^{(i)}\ \tilde{E}_\eta\ u$, $\psi_{\eta+1}(v^{(i)}) = \psi_{\eta+1}(u)$, and $v^{(i)}{}_{|m} = u_{|m}$. So $v^{(i)}\ \tilde{E}_{\xi_{i+1}}\ v^{(i+1)}$, and since $J_{\eta+1}(v^{(i)}) = J_{\xi_i}(v^{(i)})$ then again by Proposition 6.2.10 b), $v^{(i)} \in V_{\xi_i}$. Moreover by the induction hypothesis $\psi_\eta(u) = \psi_\eta(w)$ so since $u\ \tilde{E}_0\ w$ then $\psi_{\eta+1}(v^{(i)}) = \psi_{\eta+1}(w)$. Also by the induction hypothesis $u_{|m_{i+1}} = w_{|m_{i+1}}$ hence $v^{(i)}{}_{|m} = u_{|m} = (u_{|m_{i+1}})_{|m} = w_{|m}$. ◇

CLAIM 3. *$v^{(i)}\ \tilde{E}_{\xi_j}\ v^{(j)}$ for all $i \leq j$.*

<u>*Proof*</u>: Again by a downward induction on i: this is obvious for $i = j$; so assume that $v^{(i)}\ \tilde{E}_{\xi_j}\ v^{(j)}$. By construction we have $v^{(i-1)}\ \tilde{E}_{\xi_i}\ v^{(i)}$, and since $i < j$, then we also have $v^{(i-1)}\ \tilde{E}_{\xi_j}\ v^{(i)}$, hence $v^{(i-1)}\ \tilde{E}_{\xi_j}\ v^{(j)}$. ◇

Finally set $v = v^{(0)}$: by construction $v \in V_{\xi_0} = V_\xi$, $v_{|m_0} = w_{|m_0} = w_0$, and $\psi_\xi(v) = \psi_\xi(w) = \psi_\xi(w_0)^\frown\langle\tau\rangle$; moreover since $m_j \in J_{\xi_j}(v)$, it follows from Claim 3 that $v_{|m_j}\ \tilde{E}_{\xi_j}\ v^{(j)}{}_{|m_j} = w_{|m_j}$.

And since $w_{|m_j} = w_j$ this finishes the proof of the theorem. □

6.3. Proof of the basic case

THEOREM 6.3.1. *Assume that $\aleph^L_{\xi+1} < \aleph_1$. For any $Y \subset 2^\omega$ in $\Sigma^0_{1+\xi+2}$, $Z \subset 2^\omega$ in Π^1_1, and $X_0 \subset 2^\omega \times 2^\omega$ in Δ^1_1, if for any compact subset of Y there exists $z \in Z$ such that $K \times \{z\} \subset X_0$, then Y has a continuous lifting in $X := X_0 \cap (2^\omega \times Z)$.*

PROOF. By classical results we can fix a sequential tree T on $2 \times \omega_1$ such that Z is the projection of $\lceil T \rceil$ on the first factor 2^ω; moreover we can choose $T \in L$.

Applying Theorem 1.7.4 to the $\mathbf{\Sigma}^0_{1+\xi+2}$ set $\tilde{Y} := Y \times 2^\omega \subset 2^\omega \times 2^\omega$, we fix $(\hat{Y}, R, \overrightarrow{R})$ a $\mathbf{\Sigma}^0_2$- representation of $\tilde{Y}$, with $\overrightarrow{R} = (R^{(\eta)})_{\eta\leq\xi}$ is a resolution family for R in *Ext*. Also applying Proposition 2.2.2 to the $\mathbf{\Sigma}^0_2$ set $\hat{Y}$ we fix a double-tree decomposition (R^+, R^-) of R satisfying the characteristic $\mathbf{\Sigma}^0_2$ property: *The strict R^--branch of any element $(s,t) \in$ Seq(2×2) precedes its strict R^+-branch.* Since $\tilde{Y}$ is in fact $\Sigma^0_{1+\xi+2}$, then Y has a "$\mathbf{\Sigma}^0_{1+\xi+2}$ code" in L, and applying the previous results (which are absolute as Π^1_1 statements) in L, we can suppose that all parameters: $\overrightarrow{R}, R, R^+, R^-$ are in L;

<u>*The game G^**</u>: Consider $\Omega :=$ Seq$(\omega_1) \setminus \{\emptyset\}$ as "labeling space", and $\emptyset$ as an "external element", so that $\Omega^* =$ Seq(ω_1).

Let G^* be the delayed labeling game in which the players choose (s_n, t_n, τ_n) as in $\mathscr{L}(\Omega^*)$ with the following additional rules (which all of them concern *Player II*):

<u>*Rules*</u>: If $u = (s, t, \theta) \in 2^{n+1} \times 2^{n+1} \times \Omega^{n+1}$ with $\theta = (\tau_m)_{m\leq n}$ satisfies $\mathscr{L}(\Omega^*)$, then u is legal in G^* iff:

(R$_1$): *(t, τ_n) is compatible with T*

(R$_2$): *If $\varphi_*(u_{|m+1})\ R^+\ \varphi_*(u)$ and $m < n$ then $\tau_m \prec \tau_n$*

Obviously G^* is obtained by label delay from an R^+ (hence an R)-labeling game.

If $(s_n)_{n\geq0}$ and $(t_n, \tau_n)_{n\geq0}$ are the moves of the players in an infinite run in G^*, we shall say that $y = \bigcup s_n \in 2^\omega$ and $z = \bigcup t_n \in 2^\omega$ are the <u>*reals constructed*</u> by *Player I* and *Player II* respectively, in this run. We shall say that this infinite run is trivial if the set $\{n : \tau_n \neq \emptyset\}$ is finite. By the choice of T, in any non-trivial run the real z constructed by *Player II* is automatically in $Z = \text{proj}(\lceil T \rceil)$.

Win condition: *Player II* wins the infinite run $(s_n, t_n, \tau_n)_{n \geq 0}$ in G^* if:

$$y \notin Y \quad or \quad \big((y,z) \subset X_0 \textit{ and the run is non-trivial}\big)$$

Notice that G^* is a slight modification of the game introduced in Remark 6.1.2, and by the same arguments we easily have:

LEMMA 6.3.2. *If Player II has a winning strategy in G^* then Y admits a continuous lifting in X.*

LEMMA 6.3.3. *If Player I has a winning strategy in G^* then Player I has a winning strategy definable in L.*

PROOF. Notice that the game G^* is definable in L and the win condition can be written as a conjunction of a Δ^1_1 condition on reals and a closed condition on $2^\omega \times \omega_1^\omega$; the conclusion follows then by absoluteness (see [**3**], Proposition 4.3). □

Plan of the rest of the proof: Since the game G^* is determined it follows from Lemma 6.3.2 and Lemma 6.3.3 above that to prove Theorem 6.3.1, it is enough to prove that if "$\aleph^L_{\xi+1} < \aleph_1$" then *Player I* has no winning strategy in G^* which is definable in L.

In fact assuming that "$\aleph^L_{\xi+1} < \aleph_1$", we shall derive from any (non necessarily winning) strategy σ for *Player I* in G^* which is definable in L, a compact set K with the following properties:

(a) *Any $y \in K \setminus Y$ can be constructed by Player I in an infinite run compatible with σ*

(b) *Any $z \in Z$ can be constructed by Player II in a non-trivial run, compatible with σ and in which Player I constructs some real in K.*

Now if we suppose that σ is winning, then it follows from the win condition and *a)* above that $K \subset Y$. Then applying the assumption of Theorem 6.3.1, we get some $z \in Z$ such that $K \times \{z\} \subset X_0$, but then the infinite run given by *b)* above is necessarily won by *Player II*, which gives a contradiction, and proves Theorem 6.3.1.

From now on, we a fix σ a strategy for Player I in the game G^*, with code in L, and assume that $\aleph^L_{\xi+1} < \aleph_1$.

Let $W(\sigma)$ denote the set of all positions in the game G^* which are compatible with σ. We shall view $W(\sigma)$ as the set of *all* positions in a game $G^*(\sigma)$ obtained by adding formally to the rules of G^* a rule imposing to *Player I* to follow the strategy σ of G^*. Since we did not change the rules for *Player II*, it is clear that the game $G^*(\sigma)$ is also an R-labeling game with delay, to which we can apply the results of the previous section .

Let $(V_\eta, \tilde{E}_\eta, E_\eta)_{\eta \leq \xi}$ be as in Definition 6.2.7 with respect to the game $G^*(\sigma)$. We recall that we view $V_0 = W(\sigma)$ as a subset of $(2 \times 2 \times \Omega^*)^{<\omega}$, which happens to be a sequential tree on $2 \times 2 \times \Omega^*$, whereas $V_\xi \subset V_0$ for $\xi > 0$ is not. We shall in fact define next a new subset V of V_ξ which will be a sequential tree on $2 \times 2 \times \Omega^*$; it will follow that the set $S = \{\varphi(v) ; \ v \in V\}$ is a sequential tree on 2×2, hence $H = \lceil S \rceil$ is a compact subset of $2^\omega \times 2^\omega$, and we shall finally define $K \subset 2^\omega$ as the projection of $H \subset 2^\omega \times 2^\omega$ on the first factor.

Notice that any infinite branch of V (for the extension relation on positions) defines an infinite run in $G^*(\sigma)$, hence an element $(y, z) \in H$. But the converse is not a priori true, and an element $(y, z) \in \lceil S \rceil$ is not a priori realized in some infinite run in $G^*(\sigma)$.

The definition of V is not straightforward and necessitates the introduction of a number of preliminary notions. Let us recall that since $\aleph^L_{\xi+1} < \aleph_1$ then by Proposition 6.2.9, for all $\eta \leq \xi$, the equivalence relation E_η has countably many classes, hence we can fix a mapping:

$$\mathbf{N}_\eta : \mathrm{Seq}(\Omega) \longrightarrow \omega \quad \text{such that:} \quad \begin{cases} 1)\ \mathbf{N}_\eta(\emptyset) = 0 \\ 2)\ \mathbf{N}_\eta(\theta) = \mathbf{N}_\eta(\theta') \implies \theta\ E_\eta\ \theta' \end{cases}$$

and set for any position $u \in W$ with $\chi(u) = (\xi_i)_{0 \leq i \leq p}$:

$$\tilde{\mathbf{N}}(u) = \max_{0 \leq i \leq p} \mathbf{N}_{\xi_i}(\psi_{\xi_i}(u))$$

DEFINITION 6.3.4. We shall say that $u \in W$ is a minimal position if:

1) $u \in V_\xi$

2) $\psi_\xi(u) = \min\{\psi_\xi(u')\ ;\ \ u' \in V_\xi \text{ and } u'\ \tilde{E}_\xi\ u\}$

The next result asserts that the set of all minimal positions is a *section* for the restriction of the equivalence relation $\tilde{E}_\xi$ to V_ξ which is in some sense compatible with J_ξ-restrictions.

LEMMA 6.3.5. *For any position $v \in V_\xi$ there exists a unique minimal position $\mu(v)$ such that $\mu(v)\ \tilde{E}_\xi\ v$. Moreover if $m \in J_\xi(v)$ and $v_{|m}$ is minimal then*

$$\mu(v)_{|m} = v_{|m}$$

PROOF. The existence of $\mu(v)$ is obvious, and the uniqueness follows from Proposition 6.2.11.

Set $u = \mu(v)$ and suppose that $v_{|m}$ is minimal for some $m \in J_\xi(v)$; then $\psi_\xi(v_{|m}) = \psi_\xi(v)_{|k}$ for some k; hence by $\tilde{E}_0$ equivalence $\psi_\xi(u_{|m}) = \psi_\xi(u)_{|k}$ also; and since $\psi_\xi(u) \leq \psi_\xi(v)$ then by the basic property of the lexicographical ordering we also have $\psi_\xi(u_{|m}) \leq \psi_\xi(v_{|m})$.

On other hand by Proposition 6.2.10 d) and c), $u_{|m} \in V_\xi$ and $u_{|m}\ \tilde{E}_\xi\ .\ v_{|m}$; and since by hypothesis $v_{|m}$ is minimal then $\psi_\xi(v_{|m}) \leq \psi_\xi(u_{|m})$.
Hence $\psi_\xi(u_{|m}) = \psi_\xi(v_{|m})$ and again by Proposition 6.2.11 a), $u_{|m} = v_{|m}$. □

DEFINITION 6.3.6. We shall say that $u \in W$ is a good position if:

1) u is a minimal position

2) $\tilde{\mathbf{N}}(u) \leq |u|$

LEMMA 6.3.7. *The set of all good positions of fixed length is finite.*

PROOF. Fix n; and consider a good position u of length n. Since $\varphi(u) \in (2\times 2)^n$ and $\mathrm{supp}(\psi(u)) \subset [0, n[$, there are only finitely many possibilities for the $\tilde{E}_0$-class of u, hence for the characteristic sequence $\tilde{\chi}(u)$ since it is uniquely determined by the $\tilde{E}_0$-class of u.

Suppose now that the $\tilde{E}_0$-class of u and consequently the characteristic sequence $\tilde{\chi}(u) = (\xi_j)_{0 < j \leq p}$ are known. Again since u is a good position, then each of the numbers $N_j = \mathbf{N}_{\xi_j}(\psi_{\xi_j}(u)) \leq n$ can take only finitely many different values. So if we suppose moreover that the sequence $(N_j)_{0<j\leq p}$ is also known, then the E_{ξ_j}–class of $\psi_{\xi_j}(u)$ is determined for all j; and this together with the $\tilde{E}_0$-class of u, determines the $\tilde{E}_\xi$ class of u (see 6.2.12); and since u is minimal then this determines u. □

Notice that an arbitrary restriction of a minimal position is not minimal, and the same remark holds for good positions.

DEFINITION 6.3.8. We shall say that a position $v \in W$ of length n is an admissible position if $v \in V_\xi$ and there exists a sequence $(u^{(j)})_{j\leq n}$ of good positions satisfying for all $j < k \leq n$:

i) $u^{(k)}\ \tilde{E}_0\ v_{|k}$

ii) $\varphi_*(v_{|j})\ R^-\ \varphi_*(v_{|k}) \implies u^{(j)} \prec u^{(k)}$

Such a sequence $(u^{(j)})_{j\leq n}$ will be called a sequence of good versions for v.

We denote by V the set of all admissible positions.

It is clear from its definition that $V \subset (2\times 2\times \Omega^*)^{<\omega}$ is closed under arbitrary restrictions, that is V is a sequential tree on $2\times 2\times\Omega^*$, and we define as announced above:

$$S = \{\varphi(v) \ :\ v \in V\}$$

which is a sequential tree on 2×2 and

$$H = \lceil S \rceil \subset 2^\omega \times 2^\omega \quad \text{and} \quad K = \pi(H) \subset 2^\omega$$

where π denotes the canonical projection of $2^\omega \times 2^\omega$ on the first factor.

LEMMA 6.3.9. *Any $y \in K\setminus Y$ can be constructed by Player I in an infinite run in $G^*(\sigma)$*

PROOF. Fix $y \in K \setminus Y$ and pick any z such that $(y, z) \in H$. By definition of H, for all $n > 0$ we can find an admissible position u_n such that $\varphi(u_n) = (y_{|n}, z_{|n})$ and then we have $\varphi(u_n) \prec \varphi(u_{n+1})$. Fix for all n, a sequence $(u_n^{(k)})_{k\leq n}$ of good versions for u_n; since $\varphi(u_n^{(n)}) = \varphi(u_n)$ we can also suppose that $u_n^{(n)} = u_n$.

Since $\mathscr{P}(\omega)$ is compact, when identified to 2^ω, we can find some infinite subset $A \subset \omega$ such that $J := \lim_{\substack{n\to\infty\\ n\in B}} \operatorname{supp} u_n$ exists. And since $u_n^{(m)}\ \tilde{E}_0\ u_{n\,|m}$, we have for all integers m : $\lim_{\substack{n\to\infty\\ n\in B}} \operatorname{supp} u_n^{(m)} = J\cap m$. By Lemma 6.3.7 the set $\{u_n^{(m)} \ :\ n\in A\}$ is finite, hence there exists an infinite subset B of A such that the sequence $(u_n^{(m)})_{n\in B}$ is eventually constant for each m, say with value v_m, and we have $\operatorname{supp} v_m = J\cap m$ for all m and $\varphi_*(v_m) = \varphi_*(u_n^{(m)})$ for $n \in B$ large enough. Since $u_n^{(m)}\ E_0\ u_n^{(m+1)}$ we have $\varphi_*(u_n^{(m)}) \prec \varphi_*(u_n^{(m+1)})$, hence $\varphi_*(v_m) \prec \varphi_*(v_{m+1}) \prec (y, z)$. So we can find an infinite set $C \subset \omega$ such that for $m < p$ in C : $\varphi_*(v_m)\ R^-\ \varphi_*(v_p)$.

Hence if $m < p$ are in C and $n \in B$ is large enough, then

$$\varphi_*(u_k^{(m)}) = \varphi_*(v_m)\ R^-\ \varphi_*(v_p) = \varphi_*(u_k^{(p)})$$

so by admissibility $v_m = u_k^{(m)} \prec u_k^{(p)} = v_p$. This proves that there is an infinite run which extends any v_m for $m \in C$, and realizes (y, z). □

LEMMA 6.3.10. *Any $z \in Z$ can be constructed by Player II in a non-trivial run in $G^*(\sigma)$ in which Player I constructs some real in K.*

PROOF. Fix $z \in Z$, pick some $\bar{\tau} \in \omega_1^\omega$ such that $(z, \bar{\tau}) \in [T]$, and set:

$$W(z) = \{v = (s, t, \theta) \in W \ : \ t \prec z\} \quad \text{and} \quad V(z) = V \cap W(z)$$

$$W(z, \bar{\tau}) = \{v = (s, t, \theta) \in W(z) \ : \ \theta_*(|\theta_*| - 1) \prec \bar{\tau}\} \quad \text{and} \quad V(z, \bar{\tau}) = V \cap W(z, \bar{\tau})$$

where by convention $\theta_*(-1) = \emptyset$. Thus positions in $V(z, \bar{\tau})$ are positions in which *Player II* is choosing for his Seq(2)-moves beginnings of z, and in which the *last* label is a beginning of $\bar{\tau}$. Notice that the last label is not necessarily chosen in the last move, since *Player II* can "pass" by choosing $\emptyset$ (the external element) after the last label.

Let h^+ denote the height function of the tree relation R^+. We shall construct two infinite sequences $(u_n)_{n>0}$ and $(v_n)_{n>0}$ in V satisfying for all $n > 0$:

(i) u_n *is a good position*

(ii) $v_n \in V(z, \bar{\tau})$, *and* $|v_n| = n$

(iii) $\forall \eta \leq \xi \ , \ \forall m \in J_\eta(v_n) \ , \ v_{n\,|m} \ \tilde{E}_\eta \ v_m$

(iv) $u_n \ \tilde{E}_\xi \ v_n$

(v) $\varphi(v_{n-1}) \prec \varphi(v_n)$

(vi) *If* $\varphi_*(v_m) \ R^+ \varphi_*(v_n)$ *then* $v_m \prec v_n$

(vii) *If* $\varphi_*(v_m) \ R^- \varphi_*(v_n)$ *then* $u_m \prec u_n$

(viii) *If* $\varphi_*(v_m) \ R^- \varphi_*(v_n)$ *and* $h^+(\varphi_*(v_n)) = 1$ *then* $u_m \prec v_n$

and moreover:

(ix) $\lim_{n \to \infty} |\varphi_*(u_n)| = \infty$

(Notice that by *(iv)* $u_n \ \tilde{E}_0 \ v_n$ hence $\varphi(u_n) = \varphi(v_n)$ and $\varphi_*(u_n) = \varphi_*(v_n)$)

<u>*Proof of the lemma from the construction*</u>: Assume that the construction is achieved; and fix some n. It follows from *iii)* and *iv)* that for all $m \leq n$, $u_m \ \tilde{E}_0 \ v_{n\,|m}$, and since the u_m s are good positions, *vii)* ensures that v_n is an admissible position with $(u_m)_{m \leq n}$ as a sequence of good versions; hence $\varphi(v_n) \in S$ for all n and since $|\varphi_*(v_n)| \leq |\varphi(v_n)|$ then by *ix)*, $(y, z) := \bigcup_n \varphi(v_n) = \bigcup_n \varphi_*(v_n) \in \lceil S \rceil = H$ so $y \in K$. Finally if we consider the *Ext*-branch $A := \{\varphi_*(v_n) \ : n \in \omega\}$ then since (R^+, R^-) is a double-tree representation in *Ext*, we can distinguish two cases:

- either $y \in Y$; so $(y, z) \in \tilde{Y}$ and A contains an infinite R^+-branch $(\varphi_*(v_{m_j}))_{j \in \omega}$. Then by condition *vi)*, $v_{k_i} \prec v_{k_j}$ for all $i < j$, and $(v_{m_j})_{j \in \omega}$ defines an infinite run in $G^*(\sigma)$ in which the players construct (y, z); moreover condition *ix)* ensures that this run is non-trivial.

- or $y \notin Y$; then similarly A contains an infinite R^--branch $(\varphi_*(v_{m_j}))_{j \in \omega}$, and by condition *vii)*, $u_{k_i} \prec u_{k_j}$ for all $i < j$, and again $(u_{m_j})_{j \in \omega}$ defines an infinite non-trivial run in $G^*(\sigma)$ in which the players construct (y, z). ◇

This proves the lemma, and we now come back to the construction of the sequence $(u_n, v_n)_{n>0}$ for which we proceed by induction.

<u>The construction</u>: For $n = 1$, let $\sigma(\emptyset)$ be the first move of *Player I* imposed by σ; we then define

$$u_1 = v_1 = (\sigma(\emptyset), z_{|1}, \langle\emptyset\rangle)$$

So $\psi_0(u_1) = \langle\emptyset\rangle$ and $\psi_\xi(u_1) = \psi_0(u_1) = \emptyset$, hence by Proposition 6.2.10 d), $u_1 \in V_\xi$. Moreover since $\emptyset$ is the least element of $\mathrm{Seq}(\Omega)$ then u_1 is a minimal position; also since $N_\xi(\emptyset) = 0$ then u_1 is a good position; hence (since $|u_1| = 1$) u_1 is an admissible position, and one easily checks that (u_1, v_1) satisfies conditions *i)* to *viii)*.

Suppose that we already constructed positions $(u_m, v_m)_{m<n}$ satisfying conditions *i)* to *viii)*. Let $\hat{u}_{n-1}$ and $\hat{v}_{n-1}$ denote the unique trivial extensions of length n in $V(z)$, of u_{n-1} and v_{n-1} respectively. It is clear that if we set $(u_n, v_n) = (\hat{u}_{n-1}, \hat{v}_{n-1})$ then conditions *i)* to *viii)*, which are preserved by trivial extensions, are automatically satisfied. However if we follow this definition forever then obviously condition *ix)* will not be satisfied. So our plan is to define u_n and v_n differently, namely with a non-trivial last move, unless this is impossible, in which case we shall define $(u_n, v_n) = (\hat{u}_{n-1}, \hat{v}_{n-1})$. More precisely we shall define a pair (u, v) of positions with non-trivial last move, satisfying *ii)* to *viii)*, and the following weaker form of *i)*:

i') u is a minimal position

and then define:

$$(u_n, v_n) = \begin{cases} (u, v) & \text{if } \tilde{\mathbf{N}}(u) \leq n \\ (\hat{u}_{n-1}, \hat{v}_{n-1}) & \text{if } \tilde{\mathbf{N}}(u) > n \end{cases}$$

which will ensure that (u_n, v_n) satisfies *i)* to *viii)*.

Notice that since $u_1 = v_1$ is a trivial move then for any position w such that $\varphi_*(u_1) \prec \varphi(w)$ we have $\varphi_*(w) \prec \varphi(w)$ (*strict* extension). In particular this will automatically hold for u_n and v_n once they are defined. It follows that $(s, t) := \varphi_*(v_n) \preceq \varphi(v_{n-1})$ is already known, and consequently $\chi(s, t) = (\xi_i)_{0 \leq i \leq p}$ also, and we can fix a sequence $(m_i)_{0 \leq i \leq p}$ of integers $< n$ such that $\varphi_*(v_{m_i})$ is the $R^{(\xi_i)}$–predecessor of $\varphi_*(v_n)$ for all $i \in [0, p]$; moreover we can, and shall, take $m_p = n - 1$. In particular:

$$(s^*, t^*) := \varphi_*(v_{m_0}) = \varphi_*(u_{m_0}) \text{ is the } R\text{–predecessor of } (s, t)$$

Finally let h^+ denote the height function relative to the tree relation R^+, and set

$$k := h^+(s, t) > 0 \quad \text{and} \quad \tau = \bar{\tau}_{|k}$$

We shall prove next that τ is a potential label of (s, t) beyond either u_{m_0} or v_{m_0}. Then applying Theorem 6.2.14 to either $(u_{m_i})_{0 \leq i \leq p}$ or $(v_{m_i})_{0 \leq i \leq p}$ will provide us $v \in V_\xi$ with the desired properties, and finally we shall define $u = \mu(v)$ which is by definition a minimal position.

To achieve this we have to distinguish two cases. Notice first that the basic property of the double-tree (R^+, R^-) given by Proposition 2.2.2 can be transcribed as follows:

$$(\star) \qquad \begin{cases} k = 1 \iff & (s^*, t^*) \text{ is the } R^-\text{–predecessor of } (s, t) \\ k > 1 \iff & (s^*, t^*) \text{ is the } R^+\text{–predecessor of } (s, t) \end{cases}$$

<u>Case 1</u>: $k = 1$

In this case τ is trivially a potential label of (s,t) beyond u_{m_0}, and by the induction hypothesis, we can apply Theorem 6.2.14 to $(u_{m_i})_{0\leq i\leq p}$ to get $v \in V_\xi$ satisfying:

$$u_{m_0} \prec v \ , \ \varphi_*(v) = (s,t) \ , \ \psi_\xi(v) = \psi_\xi(u_{m_0})^\frown\langle\tau\rangle \ \text{ and } \ v_{|m_i} \ \tilde{E}_{\xi_i} \ u_{m_i}$$

which ensures conditions *ii)* to *viii)* (notice that in this case condition *vii)* is vacuous). Finally by Lemma 6.3.5, we have $u_{m_0} \prec u = \mu(v)$ which ensures *vii)*.

Case 2: $k > 1$

In this case (s^*, t^*) is the R^+–predecessor of (s,t), hence $h^+(s^*,t^*) = k-1$ and clearly τ is a potential label of (s,t) beyond v_{m_0}, and again by the induction hypothesis, we can apply Theorem 6.2.14 to $(v_{m_i})_{0\leq i\leq p}$ to get $v \in V_\xi$ satisfying:

$$v_{m_0} \prec v \ , \ \varphi_*(v) = (s,t) \ , \ \psi_\xi(v) = \psi_\xi(v_{m_0})^\frown\langle\tau\rangle \ \text{ and } \ v_{|m_i} \ \tilde{E}_{\xi_i} \ v_{m_i}$$

which ensures conditions *ii)* to *viii)* (in this case condition *viii)* is vacuous).

Let us now show *vii)*: notice that by the basic property of the double-tree (R^+, R^-), the R^-–predecessor of (s,t) precedes all the non-zero R^+-predecessors of (s,t); hence we can find $l < m \leq m_0$ such that:

- $\varphi_*(u_l)$ is the R^--predecessor of $\varphi_*(v_n)$
- $\varphi_*(v_m) \ R^+ \ \varphi_*(v_{m_0}) \ R^+ \ \varphi_*(v_n)$ and $h^+(\varphi_*(v_m)) = 1$

then by conditions *vii)* and *viii)* of the induction hypothesis we have $u_l \prec v_m \preceq v_{m_0} \prec v$ and as above by Lemma 6.3.5 we have $u_{m_0} \prec u = \mu(v)$ from which condition *vii)* follows.

This finishes the inductive construction of (u_n, v_n) satisfying conditions *i)* to *viii)*.

We now show that the global condition *ix)* is also ensured: for otherwise, since $\tilde{\mathbf{N}}$ is invariant by trivial extensions, then we would have for all $n > n_0$, $(u_n, v_n) = (\hat{u}_{n-1}, \hat{v}_{n-1})$ hence going back to definitions:

$$n < \tilde{\mathbf{N}}(u_n) = \tilde{\mathbf{N}}(\hat{u}_{n-1}) = \tilde{\mathbf{N}}(u_{n-1}) = \cdots = \tilde{\mathbf{N}}(u_{n_0})$$

which is impossible.

This proves condition *ix)* and ends the proof of Lemma 6.3.10, and consequently of Theorem 6.3.1 . □

6.4. Proof of the general limit case

THEOREM 6.4.1. *Assume that ξ is limit and $\aleph_\xi^L < \aleph_1$. For any $Y \subset 2^\omega$ in $\Sigma^0_{1+\xi+1}$, $Z \subset 2^\omega$ in Π^1_1, and $X_0 \subset 2^\omega \times 2^\omega$ in Δ^1_1, if for any compact subset K of Y there exists $z \in Z$ such that $K \times \{z\} \subset X_0$ then Y has a continuous lifting in $X := X_0 \cap (2^\omega \times Z)$.*

PROOF. By 4.4.3 we can assume that ξ is infinite.

Fix T as in the proof of Theorem 6.3.1, and apply Theorem 1.7.4 to the $\mathbf{\Sigma}^0_{1+\xi+1}$ set $\tilde{Y} := Y \times 2^\omega \subset 2^\omega \times 2^\omega$ to fix a ξ-distinguished subtree R of *Ext* such that $\tilde{Y}$ is the expansion of some $\mathbf{\Sigma}^0_2$ subset $\hat{\tilde{Y}}$ of $[R]$. Since ξ is limit we can now fix a *uniform*

ξ-distinguished resolution $\overrightarrow{R} = (R^{(\eta)})_{\eta \leq \xi}$ of R in *Ext*, and a double-tree decomposition (R^+, R^-) of R given by Proposition 2.2.2, as in Theorem 6.3.1. Again by general absoluteness arguments we can suppose that all parameters: $(\overrightarrow{R}, R, R^+, R^-)$ are in L.

Fix a fundamental sequence $(\eta_k)_{k\in\omega}$ such that for all (s,t), $(s',t') \in \text{Seq}(2) \times \text{Seq}(2)$ if $\min(h_R(s,t), h_R(s',t')) \leq k$ then:

$$(s,t)\, R^{(\eta_k)}\, (s',t') \implies (s,t)\, R\, (s',t')$$

and consider the family $(V_\eta, \tilde{E}_\eta, E_\eta)_{\eta\leq\xi}$ as in Definition 6.2.7. Keeping V_ξ unchanged we shall replace the equivalence relations E_ξ and $\tilde{E}_\xi$, by slightly modified versions $E_\xi^{(\mathrm{u})}$ and $\tilde{E}_\xi^{(u)}$, that we define below. Let's recall the definition of E_ξ:

$$\theta\, E_\xi\, \theta' \iff |\theta| = |\theta'| = p, \quad \text{and} \quad \forall\, k \leq p\,,\ \forall \eta < \xi\,,\ \theta_{|_k}\, E_\eta\, \theta'_{\ |_k}$$

We now define:

$$\theta\, E_\xi^{(\mathrm{u})}\, \theta' \iff |\theta| = |\theta'| = p, et\ \forall\, k \leq p\,,\ \theta_{|_k}\, E_{\eta_k}\, \theta'_{\ |_k}$$

and

$$u\, \tilde{E}_\xi^{(\mathrm{u})}\, u' \iff \forall \eta < \xi,\ u\, \tilde{E}_\eta\, u', \quad \text{and} \quad \psi_\xi(u)\, E_\xi^{(\mathrm{u})}\, \psi_\xi(u')$$

LEMMA 6.4.2. *The equivalence relations $E_\xi^{(\mathrm{u})}$ and $\tilde{E}_\xi^{(\mathrm{u})}$ have at most $\aleph_\xi^L$ classes.*

We leave the immediate proof of the lemma to the reader.

We now have to modify the notion of characteristic sequence. For this we proceed as in 6.2.12 in two steps:

– In the "abstract setting" (see 6.2.12) the <u>uniform characteristic sequence</u> of an element a is the sequence $\chi^{(\mathrm{u})}(a) := (\xi_i)_{0\leq i\leq n}$ defined by:

$$\begin{cases} \xi_0 = \eta_k & \text{if } h_R(a) = k \\ \xi_i = \max\{\zeta \leq \xi_{i-1} : A_\zeta \neq A_{\xi_{i-1}}\} & \text{for } i > 0 \end{cases}$$

– In the "game setting" we define again for any position u:

$$\bar{\chi}^{(\mathrm{u})}(u) = \chi^{(\mathrm{u})}(\varphi_*(u))$$

One immediately checks that the basic property pointed to in paragraph 6.2.12 holds:

$$u\, E_\xi^{(\mathrm{u})}\, v \iff \begin{cases} u\, \tilde{E}_0\, v & \left(\text{with } \bar{\chi}^{(\mathrm{u})}(u) = \bar{\chi}^{(\mathrm{u})}(v) = (\xi_i)_{0\leq i\leq n}\right) \\ \text{and} & \\ \psi_{\xi_i}(u)\, E_{\xi_i}^{(\mathrm{u})}\, \psi_{\xi_i}(v) & \text{for all } i \in [0,n] \end{cases}$$

The rest of the proof is then obtained by replacing systematically in the arguments of Section 6.2 E_ξ, $\tilde{E}_\xi$, χ by their "uniform" versions $E_\xi^{(\mathrm{u})}$, $\tilde{E}_\xi^{(\mathrm{u})}$, $\chi^{(\mathrm{u})}$. □

6.5. Proof of the general successor case

THEOREM 6.5.1. *Assume that* $\aleph_{\xi+1}^L < \aleph_1$. *For any* $Y \subset 2^\omega$ *in* $D(\Sigma^0_{1+\xi+2})$, $Z \subset 2^\omega$ *in* Π^1_1, *and* $X_0 \subset 2^\omega \times 2^\omega$ *in* Δ^1_1, *if for any compact subset* K *of* Y *there exists* $z \in Z$ *such that* $K \times \{z\} \subset X_0$ *then* Y *has a continuous lifting in* $X := X_0 \cap (2^\omega \times Z)$.

PROOF. Consider as in the proof of Theorem 6.3.1, $\tilde{Y} = Y \times 2^\omega$, and a ξ-distinguished subtree R of *Ext* such that $\tilde{Y}$ is the expansion of some $D(\mathbf{\Sigma}^0_2)$ subset $\widehat{\tilde{Y}}$ of $[R]$. Fix then a ξ-distinguished resolution $\overrightarrow{R} = (R^{(\eta)})_{\eta \leq \xi}$ of R in *Ext*, and a double-tree representation (R^+, R^-) of R of $\widehat{\tilde{Y}}$ given by Proposition 2.3.1, that is such that: *the strict* R^+*-chain of any element is entirely between two consecutive elements of the* R^-*-chain of this element.* Fix also T as in Theorem 6.3.1, and set $\Omega = \mathrm{Seq}(\omega)_1 \setminus \{\emptyset\}$.

We shall introduce a game which is a variation of the game G^* considered in the proof of Theorem 6.3.1.

<u>The game G^{**}</u>: This is also a delayed labeling game in which the players choose (s_n, t_n, τ_n) as in G^* following rules (R_1), (R_2) as in 6.3.1 and the following additional rule:

(R_3): *For all* $m < |u_*|-1$ *and all* $n < |u_*|-1$, *if* $\varphi_*(u)_{|m+1}$ *and* $\varphi_*(u)_{|n+1}$ *are of* R^+*–height* > 1 *and have the same* R^+*–predecessor then* $\psi_*(u)(m) = \psi_*(u)(n)$.

We keep the win condition unchanged: If (y, z) is the pair of reals constructed by the players in some infinite run in G^{**}, *Player II* wins the run if:

$$y \notin Y \quad \text{or} \quad \big((y, z) \in X_0 \text{ and the run is non-trivial}\big)$$

Obviously Lemma 6.3.2 and Lemma 6.3.3 are also valid for G^{**}. We also fix a strategy σ for *Player I* definable in L, and consider the families $(V_\eta, \tilde{E}_\eta, E_\eta)_{\eta \leq \xi}$ obtained by applying the results of Section 6.2 to $G^{**}(\sigma)$. Then assuming that $\aleph_{\xi+1}^L < \aleph_1$ we fix mappings $(\tilde{\mathbf{N}}_\eta)_{\eta \leq \xi}$ and $\tilde{\mathbf{N}}$ as before and construct (V, S, H, K) exactly with the same definitions from $(V_\eta, \tilde{E}\text{ \_eta}, E_\eta)_{\eta \leq \xi}$. Again Lemmas 6.3.5, 6.3.7, 6.3.9 are still valid, with the same proofs.

However Lemma 6.3.10 used among its arguments, namely in $(\star)$, the characteristic $\mathbf{\Sigma}^0_2$ property of the double-tree given by Proposition 2.2.2, which does no more hold now. Consequently the proof of this result in the new context of the game G^{**} and the double-tree given by Proposition 2.3.1 has to be suitably modified. This modification will rely on the following general property (which is also valid, but useless, in the context of G^*).

LEMMA 6.5.2. *Let* $\theta, \theta' \in \Omega^k$ *and* $i < k$. *If* $\theta \; E_\eta \; \theta'$ *then* $\theta^\frown\langle\theta(i)\rangle \; E_\eta \; \theta'^\frown\langle\theta'(i)\rangle$.

PROOF. By induction on η: the result is trivial for $\eta = 0$; and the limit case is straightforward, since for λ limit, $E_\lambda = \bigcap_{\zeta<\lambda} E_\zeta$.

Assume that the result is true for η. Let θ, $\theta' \in \Omega^k$ and $i < k$, and suppose that $\theta \; E_{\eta+1} \; \theta'$. We shall first prove that $\theta^\frown\langle\theta(i)\rangle \; F_{\eta+1} \; \theta'^\frown\langle\theta'(i)\rangle$ (with the notations of Definition 6.2.7): so consider $u \in V_{\eta+1}$ such that $\theta^\frown\langle\theta(i)\rangle \preceq \psi_{\eta+1}(u)$; then we can find some $m \in J_{\eta+1}(u)$ such that $\theta = \psi_{\eta+1}(u)_{|m}$ and $\psi_{\eta+1}(u)(m) = \theta(i)$. Since $\theta \; E_{\eta+1} \; \theta'$ then by Lemma 6.2.15, we can find $u' \in V_{\eta+1}$ such that $\theta' \preceq \psi_{\eta+1}(u')$

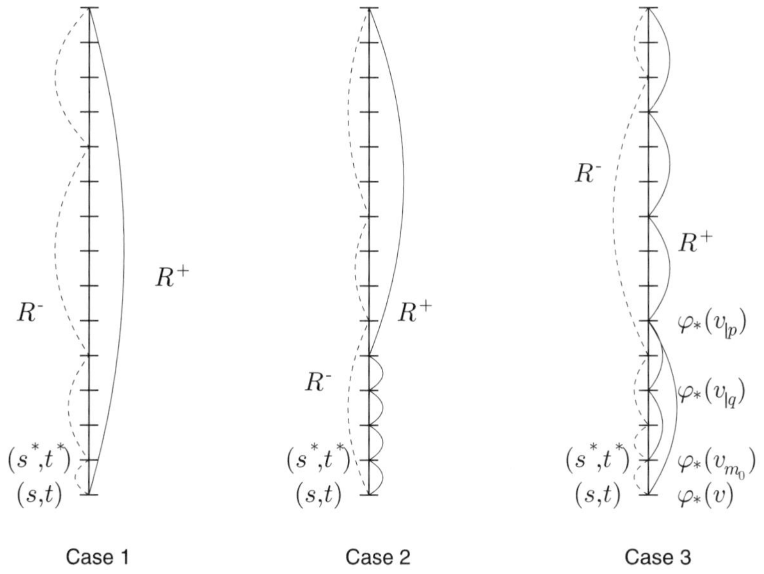

FIGURE 6.3

and $u\ \tilde{E}_\eta\ u'$. In particular $u\ \tilde{E}_0\ u'$ and $\psi(u)\ E_0\ \psi(u')$; and it follows from the definition of the equivalence relation E_0 that necessarily $\psi_{\eta+1}(u')(m) = \theta'(i)$ hence $\theta'^\frown\langle\theta'(i)\rangle \preceq \psi_{\eta+1}(u')$ with $u' \in V_{\eta+1}$ and $u\ \tilde{E}_\eta\ u'$, which proves that $\theta^\frown\langle\theta(i)\rangle\ F_{\eta+1}\ \theta'^\frown\langle\theta'(i)\rangle$. Then by symmetry we also have $\theta'^\frown\langle\theta'(i)\rangle\ F_{\eta+1}\ \theta^\frown\langle\theta(i)\rangle$, and since by hypothesis $\theta\ E_{\eta+1}\ \theta'$ then it follows from the definitions that $\theta^\frown\langle\theta(i)\rangle\ E_{\eta+1}\ \theta'^\frown\langle\theta'(i)\rangle$. □

To finish the proof of Theorem 6.5.1 we indicate the modifications to bring to the proof of Lemma 6.3.10: we follow exactly the same scheme, in particular constructing, by induction, an infinite sequence $(u_n, v_n)_{n>0}$ satisfying the same conditions *i)* to *ix)*. The definition of (u_n, v_n) is also of the form

$$(u_n, v_n) = \begin{cases} (u, v) & \text{if } \tilde{\mathbf{N}}(u) \leq n \\ \\ (\hat{u}_{n-1}, \hat{v}_{n-1}) & \text{if } \tilde{\mathbf{N}}(u) > n \end{cases}$$

The main new difficulty is that the equivalence $(\star)$ does not hold any more; instead, we only have:

$$(\star\star) \qquad \begin{cases} k = 1 \implies & (s^*, t^*) \textit{ is the } R^-\textit{–predecessor of } (s,t) \\ k > 1 \impliedby & (s^*, t^*) \textit{ is the } R^+\textit{–predecessor of } (s,t) \end{cases}$$

and for the construction of (u, v) one now has to distinguish three cases:

- *Case 1:* $k = 1$
- *Case 2:* $k > 1$ *and the R–predecessor of* (s,t) *is the* R^+*–predecessor*
- *Case 3:* $k > 1$ *and the R–predecessor of* (s,t) *is the* R^-*–predecessor*

where, as in the basic case, k denotes the height $h^+(s,t)$ of (s,t) with respect to R^+.

Notice that in Cases 1 and 2, rule (R_3) is vacuous; moreover the R^+–predecessor and the R^-–predecessor of (s,t) are in the same configuration as in the game G^*, and one can repeat exactly the same arguments as in G^{**}.

In Case 3, observe that (s^*,t^*) is the R^-–predecessor of (s,t), and we can find $0 < p < m_0$ such that $\varphi_*(v_p)$ is the R^+–predecessor of (s,t), hence by the distinction of R^+ in R we have: $\varphi_*(v_p)\ R^+\ \varphi_*(v_{m_0})\ R^-\ (s,t)$. If $q \in [p, m_0]$ is such that $\varphi_*(v_p)$ is the R^+–predecessor of $\varphi_*(v_q)$, then by condition *vi)* of the induction hypothesis we have $v_p \prec v_q \preceq v_{m_0}$ hence by rule (R_3) the last labeling in v_q is necessarily $\tau = \bar{\tau}_{|k}$. This shows that τ is a potential label of (s,t) beyond v_{m_0}, and again applying Theorem 6.2.14 to $(v_{m_i})_{0\leq i\leq p}$ we get $v \in V_\xi$ satisfying conditions *ii)* to *vi)* as well as *viii)*.

Finally as in G^* to ensure condition *vii)* we have to prove that $u_{m_0} \prec u = \mu(v)$. For this observe that $\psi_\xi(v) = \psi_\xi(v_{m_0})^\frown\psi_\xi(v_{m_0})(i)$ for some $i < |\psi_\xi(v_{m_0})|$, and consider $\tau' = \psi_\xi(u_{m_0})(i)$. Then clearly τ' is a potential label of (s,t) beyond u_{m_0}, and we can once again apply Theorem 6.2.14 to now $(u_{m_i})_{0\leq i\leq p}$ to get some $u' \in V_\xi$, necessarily with $\psi_\xi(u') = \psi_\xi(u_{m_0})^\frown\psi_\xi(u_{m_0})(i)$. But by the induction hypothesis we have $\psi_\xi(u_{m_0})\ E_\xi\ \psi_\xi(v_{m_0}) = \psi_\xi(v_{|m_0})$, hence by Lemma 6.5.2 $\psi_\xi(u')\ E_\xi\ \psi_\xi(v)$ and so $u'\ \tilde{E}_\xi\ v$. Moreover since u_{m_0} is minimal u' is automatically minimal, hence $u' = \mu(v) \succ u_{m_0}$.

The rest of the proof goes exactly in the same way as for G^*. □

REMARKS 6.5.3. a) If the double-tree (R^+, R^-) is given by Proposition 2.2.2 then rule (R_3) is vacuous, and in this case the game G^{**} coincides with the game G^*.

b) As we mentioned before, because of Corollary 5.3.5, one cannot extend the previous result to the limit case. Still one could be tempted to replace the last equivalence relations E_ξ and $\tilde{E}_\xi$ in the proof of Theorem 6.5.1 by their uniform versions $E_\xi^{(\mathrm{u})}$ and $\tilde{E}_\xi^{(\mathrm{u})}$, as in the proof of Theorem 6.4.1. The problem arises then when proving Lemma 6.5.2 precisely for the *final* value $\eta = \xi$ which is a limit ordinal but for which $E_\xi^{(\mathrm{u})}$ is no more $\bigcap_{\zeta<\xi} E_\zeta$.

Bibliography

1. J.P.R. Christensen, *Necessary and sufficient conditions for the measurability of certain sets of closed sets*, Math. Annalen **200** (1973), 189–193
2. G. Debs and J. Saint Raymond, *Compact covering and game determinacy*, Topology Appl.**68** (1996), 153–185
3. G. Debs and J. Saint Raymond, *Cofinal* $\mathbf{\Sigma}_1^1$ *and* $\mathbf{\Pi}_1^1$ *subsets of* ω^ω, Fund. Math.**159** (1999), 161–193
4. G. Debs and J. Saint Raymond, *Compact covering mappings and cofinal families of compact subsets of a Borel set*, Fundamenta Math. **167** (2001), 213–249
5. G. Debs and J. Saint Raymond, *Applications semi-propres sur un espace borélien*, C.R.A.S. Paris, Série I, **t.332** (2001), 423–426
6. G. Debs and J. Saint Raymond, *Compact covering mappings between Borel sets and the size of constructible reals*, Transactions of the Amer. Math. Soc. **356** (2004), no. 1, 73-117
7. G. Debs and J. Saint Raymond, *Arbres distingués, bi-arbres et théorèmes de relèvement*, C.R.A.S. Paris, Série I, **t.336** (2003), 625–628
8. H. Friedman, *Selection for Borel Relations*, The Mathematics Preprint Server (http://www.mathpreprints.com) (2001)
9. L. Harrington, *Analytic Determinacy and* $0^\sharp$, J. Symboloic Logic **43** (1978), 685–693
10. W. Just and H. Wicke, *Some conditions under which tri-quotient or compact-covering maps are inductively perfect*, Topology Appl. **55** (1994), 289–305
11. A. Kechris, *Classical Descriptive Set Theory*, (Springer-Verlag, New York, 1995)
12. Kuratowski, *Topology vol. I* , (Academic Press, New York, 1966)
13. A. Louveau, *Some results in the Wadge Hierarchy of Borel sets*, CABAL Seminar 79-81, Lecture Notes in Math. **1019**, Springer-Verlag, 1983
14. A. Louveau and J. Saint Raymond, *Borel classes and closed games : Wadge-type and Hurewicz-type results*, Trans. Amer. Math. Soc. **304** vol. 2 (1987) 431-467
15. A. Louveau and J. Saint Raymond, *The strength of Borel Wadge determinacy*, CABAL Seminar 81-85, Lecture Notes in Math. **1333**, 1-30
16. N. Lusin, *Leçons sur les ensembles analytiques et leurs applications*, (Chelsea Publishing Company, New York, 1972)
17. D.A. Martin, *Measurable cardinals and analytic games*, Fund. Math. **66** (1969/1970), 287–291
18. Y.N. Moschovakis, *Descriptive Set Theory*, (North Holland, Amsterdam, 1980)
19. A.V. Ostrovsky, *On compact-covering and open maps of Borel sets* Séminaire d'Initiation à l'Analyse, 32ème année, n° 17 (1992-93)
20. A.V. Ostrovsky, *New class of maps connected with compact covering maps*, Vestnik Moskov. Gos. Univ. **1994** n° 4, 24–27
21. J. Saint Raymond, *Caractérisation d'espaces polonais*, Séminaire d'Initiation à l'Analyse, 11–12èmes années, n° 5 (1971-73)
22. J. Saint Raymond, *Fonctions boréliennes sur un quotient*, Bull. Sc. Math. **100** (1976), 141-147
23. R. L. Sami, *On the equivalence relations with Borel classes of bounded rank*, J. of Symb. Logic **49** (1984), 1273–1283
24. R.M. Solovay, *On the cardinality of* Σ_2^1 *sets of reals*, in : Foundations of Mathematics (Symposium commemorating Kurt Gödel, Columbus, OH, 1996), Springer, New York, 1969, 58–73
25. J. Stern, *Effective partitions of the real line into Borel sets of bounded rank*, Annals of Mathematical Logic **18** (1980), 29–60
26. J. Stern, *On Lusins's restricted continuum problem*, Annals of Math. **120** (1984), 7–33
27. W.W. Wadge, *Thesis*, Berkeley, 1984

Index

Editorial Information

To be published in the *Memoirs*, a paper must be correct, new, nontrivial, and significant. Further, it must be well written and of interest to a substantial number of mathematicians. Piecemeal results, such as an inconclusive step toward an unproved major theorem or a minor variation on a known result, are in general not acceptable for publication.

Papers appearing in *Memoirs* are generally at least 80 and not more than 200 published pages in length. Papers less than 80 or more than 200 published pages require the approval of the Managing Editor of the Transactions/Memoirs Editorial Board.

As of February 28, 2007, the backlog for this journal was approximately 15 volumes. This estimate is the result of dividing the number of manuscripts for this journal in the Providence office that have not yet gone to the printer on the above date by the average number of monographs per volume over the previous twelve months, reduced by the number of volumes published in four months (the time necessary for preparing a volume for the printer). (There are 6 volumes per year, each usually containing at least 4 numbers.)

A Consent to Publish and Copyright Agreement is required before a paper will be published in the *Memoirs*. After a paper is accepted for publication, the Providence office will send a Consent to Publish and Copyright Agreement to all authors of the paper. By submitting a paper to the *Memoirs*, authors certify that the results have not been submitted to nor are they under consideration for publication by another journal, conference proceedings, or similar publication.

Information for Authors

Memoirs are printed from camera copy fully prepared by the author. This means that the finished book will look exactly like the copy submitted.

Initial submission. The AMS uses Centralized Manuscript Processing for initial submissions. Authors should submit a PDF file using the Initial Manuscript Submission form found at `www.ams.org/cgi-bin/peertrack/submission.pl`, or send one copy of the manuscript to the following address: Centralized Manuscript Processing, MEMOIRS OF THE AMS, 201 Charles Street, Providence, RI 02904-2294 USA. If a paper copy is being forwarded to the AMS, indicate that it is for it Memoirs and include the name of the corresponding author, contact information such as email address or mailing address, and the name of an appropriate Editor to review the paper (see the list of Editors below).

The paper must contain a *descriptive title* and an *abstract* that summarizes the article in language suitable for workers in the general field (algebra, analysis, etc.). The *descriptive title* should be short, but informative; useless or vague phrases such as "some remarks about" or "concerning" should be avoided. The *abstract* should be at least one complete sentence, and at most 300 words. Included with the footnotes to the paper should be the 2000 *Mathematics Subject Classification* representing the primary and secondary subjects of the article. The classifications are accessible from `www.ams.org/msc/`. The list of classifications is also available in print starting with the 1999 annual index of *Mathematical Reviews*. The Mathematics Subject Classification footnote may be followed by a list of *key words and phrases* describing the subject matter of the article and taken from it. Journal abbreviations used in bibliographies are listed in the latest *Mathematical Reviews* annual index. The series abbreviations are also accessible from `www.ams.org/publications/`. To help in preparing and verifying references, the AMS offers MR Lookup, a Reference Tool for Linking, at `www.ams.org/mrlookup/`.

Electronically prepared manuscripts. The AMS encourages electronically prepared manuscripts, with a strong preference for $\mathcal{A}_\mathcal{M}\mathcal{S}$-LaTeX. To this end, the Society has prepared $\mathcal{A}_\mathcal{M}\mathcal{S}$-LaTeX author packages for each AMS publication. Author packages include instructions for preparing electronic manuscripts, samples, and a style file that generates the particular design specifications of that publication series. Though $\mathcal{A}_\mathcal{M}\mathcal{S}$-LaTeX is the highly preferred format of TeX, author packages are also available in $\mathcal{A}_\mathcal{M}\mathcal{S}$-TeX.

Authors may retrieve an author package from the AMS website starting from `www.ams.org/tex/` or via FTP to `ftp.ams.org` (login as `anonymous`, enter username as password, and type `cd pub/author-info`). The *AMS Author Handbook* and the *Instruction Manual* are available in PDF format following the author packages link from `www.ams.org/tex/`. The author package can also be obtained free of charge by sending email to `tech-support@ams.org` (Internet) or from the Publication Division, American Mathematical Society, 201 Charles St., Providence, RI 02904-2294, USA. When requesting an author package, please specify $\mathcal{A}_\mathcal{M}\mathcal{S}$-LaTeX or $\mathcal{A}_\mathcal{M}\mathcal{S}$-TeX and the publication in which your paper will appear. Please be sure to include your complete mailing address.

After acceptance. The final version of the electronic file should be sent to the Providence office (this includes any TeX source file, any graphics files, and the DVI or PostScript file) immediately after the paper has been accepted for publication.

Before sending the source file, be sure you have proofread your paper carefully. The files you send must be the EXACT files used to generate the proof copy that was accepted for publication. For all publications, authors are required to send a printed copy of their paper, which exactly matches the copy approved for publication, along with any graphics that will appear in the paper.

Accepted electronically prepared files can be submitted via the web at `www.ams.org/submit-book-journal/`, sent via FTP, or sent on CD-Rom or diskette to the Electronic Prepress Department, American Mathematical Society, 201 Charles Street, Providence, RI 02904-2294 USA. TeX source files, DVI files, and PostScript files can be transferred over the Internet by FTP to the Internet node `ftp.ams.org` (130.44.1.100). When sending a manuscript electronically via CD-Rom or diskette, please be sure to include a message identifying the paper as a Memoir.

Electronically prepared manuscripts can also be sent via email to `pub-submit@ams.org` (Internet). In order to send files via email, they must be encoded properly. (DVI files are binary and PostScript files tend to be very large.)

Electronic graphics. Comprehensive instructions on preparing graphics are available at `www.ams.org/jourhtml/`. A few of the major requirements are given here.

Submit files for graphics as EPS (Encapsulated PostScript) files. This includes graphics originated via a graphics application as well as scanned photographs or other computer-generated images. If this is not possible, TIFF files are acceptable as long as they can be opened in Adobe Photoshop or Illustrator. No matter what method was used to produce the graphic, it is necessary to provide a paper copy to the AMS.

Authors using graphics packages for the creation of electronic art should also avoid the use of any lines thinner than 0.5 points in width. Many graphics packages allow the user to specify a "hairline" for a very thin line. Hairlines often look acceptable when proofed on a typical laser printer. However, when produced on a high-resolution laser imagesetter, hairlines become nearly invisible and will be lost entirely in the final printing process.

Screens should be set to values between 15% and 85%. Screens which fall outside of this range are too light or too dark to print correctly. Variations of screens within a graphic should be no less than 10%.

Inquiries. Any inquiries concerning a paper that has been accepted for publication should be sent to `memo-query@ams.org` or directly to the Electronic Prepress Department, American Mathematical Society, 201 Charles St., Providence, RI 02904-2294 USA.

Editors

This journal is designed particularly for long research papers, normally at least 80 pages in length, and groups of cognate papers in pure and applied mathematics. Papers intended for publication in the *Memoirs* should be addressed to one of the following editors. The AMS uses Centralized Manuscript Processing for initial submissions to AMS journals. Authors should follow instructions listed on the Initial Submission page found at www.ams.org/memo/memosubmit.html.

Algebra to ALEXANDER KLESHCHEV, Department of Mathematics, University of Oregon, Eugene, OR 97403-1222; email: ams@noether.uoregon.edu

Algebra and its application to MINA TEICHER, Emmy Noether Research Institute for Mathematics, Bar-Ilan University, Ramat-Gan 52900, Israel; email: teicher@macs.biu.ac.il

Algebraic geometry to DAN ABRAMOVICH, Department of Mathematics, Brown University, Box 1917, Providence, RI 02912; email: amsedit@math.brown.edu

Algebraic number theory to V. KUMAR MURTY, Department of Mathematics, University of Toronto, 100 St. George Street, Toronto, ON M5S 1A1, Canada; email: murty@math.toronto.edu

Algebraic topology to ALEJANDRO ADEM, Department of Mathematics, University of British Columbia, Room 121, 1984 Mathematics Road, Vancouver, British Columbia, Canada V6T 1Z2; email: adem@math.ubc.ca

Combinatorics to JOHN R. STEMBRIDGE, Department of Mathematics, University of Michigan, Ann Arbor, Michigan 48109-1109; email: JRS@umich.edu

Complex analysis and harmonic analysis to ALEXANDER NAGEL, Department of Mathematics, University of Wisconsin, 480 Lincoln Drive, Madison, WI 53706-1313; email: nagel@math.wisc.edu

Differential geometry and global analysis to LISA C. JEFFREY, Department of Mathematics, University of Toronto, 100 St. George St., Toronto, ON Canada M5S 3G3; email: jeffrey@math.toronto.edu

Dynamical systems and ergodic theory to AMIE WILKINSON, Department of Mathematics, Northwestern University, 2033 Sheridan Road, Evanston, IL 60208-2730; email: transactions@math.northwestern.edu

Functional analysis and operator algebras to DIMITRI SHLYAKHTENKO, Department of Mathematics, University of California, Los Angeles, CA 90095; email: shlyakht@math.ucla.edu

Geometric analysis to WILLIAM P. MINICOZZI II, Department of Mathematics, Johns Hopkins University, 3400 N. Charles St., Baltimore, MD 21218; email: trans@math.jhu.edu

Geometric analysis to MLADEN BESTVINA, Department of Mathematics, University of Utah, 155 South 1400 East, JWB 233, Salt Lake City, Utah 84112-0090; email: bestvina@math.utah.edu

Harmonic analysis, representation theory, and Lie theory to ROBERT J. STANTON, Department of Mathematics, The Ohio State University, 231 West 18th Avenue, Columbus, OH 43210-1174; email: stanton@math.ohio-state.edu

Logic to STEFFEN LEMPP, Department of Mathematics, University of Wisconsin, 480 Lincoln Drive, Madison, Wisconsin 53706-1388; email: lempp@math.wisc.edu

Partial differential equations to GUSTAVO PONCE, Department of Mathematics, South Hall, Room 6607, University of California, Santa Barbara, CA 93106; email: ponce@math.ucsb.edu

Partial differential equations and dynamical systems to PETER POLACIK, School of Mathematics, University of Minnesota, Minneapolis, MN 55455; email: polacik@math.umn.edu

Probability and statistics to KRZYSZTOF BURDZY, Department of Mathematics, University of Washington, Box 354350, Seattle, Washington 98195-4350; email: burdzy@math.washington.edu

Real analysis and partial differential equations to DANIEL TATARU, Department of Mathematics, University of California, Berkeley, Berkeley, CA 94720; email: tataru@math.berkeley.edu

All other communications to the editors should be addressed to the Managing Editor, ROBERT GURALNICK, Department of Mathematics, University of Southern California, Los Angeles, CA 90089-1113; email: guralnic@math.usc.edu.

Titles in This Series

879 **O. García-Prada, P. B. Gothen, and V. Muñoz,** Betti numbers of the moduli space of rank 3 parabolic Higgs bundles, 2007

878 **Alessandra Celletti and Luigi Chierchia,** KAM stability and celestial mechanics, 2007

877 **María J. Carro, José A. Raposo, and Javier Soria,** Recent developments in the theory of Lorentz spaces and weighted inequalities, 2007

876 **Gabriel Debs and Jean Saint Raymond,** Borel liftings of Borel sets: Some decidable and undecidable statements, 2007

875 **C. Krattenthaler and T. Rivoal,** Hypergéométrie et fonction zêta de Riemann, 2007

874 **Sonia Natale,** Semisolvability of semisimple Hopf algebras of low dimension, 2007

873 **A. J. Duncan,** Exponential genus problems in one-relator products of groups, 2007

872 **Anthony V. Geramita, Tadahito Harima, Juan C. Migliore, and Yong Su Shin,** The Hilbert function of a level algebra, 2007

871 **Pascal Auscher,** On necessary and sufficient conditions for L^p-estimates of Riesz transforms associated to elliptic operators on $\mathbb{R}^n$ and related estimates, 2007

870 **Takuro Mochizuki,** Asymptotic behaviour of tame harmonic bundles and an application to pure twistor D-modules, Part 2, 2007

869 **Takuro Mochizuki,** Asymptotic behaviour of tame harmonic bundles and an application to pure twistor D-modules, Part 1, 2007

868 **Gelu Popescu,** Entropy and multivariable interpolation, 2006

867 **Vilmos Totik,** Metric properties of harmonic measures, 2006

866 **William Craig,** Semigroups underlying first-order logic, 2006

865 **Nathanial P. Brown,** Invariant means and finite representation theory of $C*$-algebras, 2006

864 **John M. Lee,** Fredholm operators and Einstein metrics on conformally compact manifolds, 2006

863 **M. Lübke and A. Teleman,** The Universal Kobayashi-Hitchin correspondence on Hermitian manifolds, 2006

862 **Alberto Canonaco,** The Beilinson complex and canonical rings of irregular surfaces, 2006

861 **Leon A. Takhtajan and Lee-Peng Teo,** Weil-Petersson metric on the universal Teichmüller space, 2006

860 **Thomas M. Fiore,** Pseudo limits, biadjoints and pseudo algebras: Categorical foundations of conformal field theory, 2006

859 **N. Arcozzi, R. Rochberg, and E. Sawyer,** Carleson measures and interpolating sequences for Besov spaces on complex balls, 2006

858 **Enrico Valdinoci, Berardino Sciunzi, and Vasile Ovidiu Savin,** Flat level set regularity of p-Laplace phase transitions, 2006

857 **Donatella Danielli, Nocola Garofalo, and Duy-Minh Nhieu,** Non-doubling Ahlfors measures, perimeter measures, and the characterization of the trace spaces of Sobolev functions in Carnot-Carathéodory spaces, 2006

856 **Vladimir Bolotnikov and Harry Dym,** On boundary interpolation for matrix valued Schur functions, 2006

855 **Yevgenia Kashina, Yorck Sommerhäuser, and Yongchang Zhu,** On higher Frobenius-Schur indicators, 2006

854 **Noam Greenberg,** The role of true finiteness in the admissible recursively enumerable degrees, 2006

853 **Joachim Krieger,** Stability of spherically symmetric wave maps, 2006

852 **Viorel Barbu, Irena Lasiecka, and Roberto Triggiani,** Tangential boundary stabilization of Navier-Stokes equations, 2006